T0257821

Autophagy: Beyond the Basics

Autophagy:
Beyond the Basics

Edited by **Alexandra Hewitt**

New York

Published by Callisto Reference,
106 Park Avenue, Suite 200,
New York, NY 10016, USA
www.callistoreference.com

Autophagy: Beyond the Basics
Edited by Alexandra Hewitt

International Standard Book Number: 978-1-63239-083-7 (Hardback)

This book contains information obtained from authentic and highly regarded sources. Copyright for all individual chapters remain with the respective authors as indicated. A wide variety of references are listed. Permission and sources are indicated; for detailed attributions, please refer to the permissions page. Reasonable efforts have been made to publish reliable data and information, but the authors, editors and publisher cannot assume any responsibility for the validity of all materials or the consequences of their use.

The publisher's policy is to use permanent paper from mills that operate a sustainable forestry policy. Furthermore, the publisher ensures that the text paper and cover boards used have met acceptable environmental accreditation standards.

Trademark Notice: Registered trademark of products or corporate names are used only for explanation and identification without intent to infringe.

Printed in the United States of America.

Contents

Preface

This book highlights some of the demanding research topics related to autophagy and also analyzes the recent developments in its molecular mechanisms. The emphasis is on the role this cell defense mechanism plays in studying various diseases which include liver diseases, cancers, myopathies and infectious diseases. The contradictory role of autophagy i.e. cell survival versus cell death; draws the focus on the importance of keeping in mind this double-edged nature in future developments of the currently promising autophagy-modulating therapies. This book provides new ground breaking researches and expands the base for further study in the field of autophagy.

The researches compiled throughout the book are authentic and of high quality, combining several disciplines and from very diverse regions from around the world. Drawing on the contributions of many researchers from diverse countries, the book's objective is to provide the readers with the latest achievements in the area of research. This book will surely be a source of knowledge to all interested and researching the field.

In the end, I would like to express my deep sense of gratitude to all the authors for meeting the set deadlines in completing and submitting their research chapters. I would also like to thank the publisher for the support offered to us throughout the course of the book. Finally, I extend my sincere thanks to my family for being a constant source of inspiration and encouragement.

Editor

Autophagy in Infectious Diseases

Autophagic Balance Between Mammals and Protozoa: A Molecular, Biochemical and Morphological Review of Apicomplexa and Trypanosomatidae Infections

Thabata Lopes Alberto Duque,

Xênia Macedo Souto,

Valter Viana de Andrade-Neto,

Vítor Ennes-Vidal and

Rubem Figueiredo Sadok Menna-Barreto

Additional information is available at the end of the chapter

1. Introduction

Protozoa are unicellular eukaryotes that are able to live as parasites or as free-living organisms and interact with a great variety of environments and organisms, from bacteria to man; in addition, they represent one of most important sources of parasitic diseases. Every year, more than one million people die from complications from protozoal infections worldwide [1-5]. Of the medically relevant protozoa, Trypanosomatidae and Apicomplexa constitute a substantial group including the causative agents of several human diseases such as Chagas disease, sleeping sickness, leishmaniasis, malaria and toxoplasmosis [1,5,6]. The life cycles of these parasites are highly complex, involving different hosts and different specific interactions with a variety of cells and tissues [7- 11]. Some of these parasites live in the extracellular matrix or blood of host mammals, but the majority of them infect host cells to complete their cycle. Despite the high infection and mortality rates of these protozoa, especially in low-income populations of developing regions such as Africa, Asia and the Americas, current therapies for these parasitic diseases are very limited and unsatisfactory. The development of efficient drugs is urgently necessary, as are serious public health initiatives to improve patients' quality of life [12-16].

The Trypanosomatidae family belongs to the order Kinetoplastida and is comprised of flagellated protists characterised by the presence of the kinetoplast, a DNA-enriched portion of the mitochondrion localised close to the flagellar pocket. The most studied pathogenic trypanosomatids are the following: (a) *Trypanosoma brucei*, which is responsible for sleeping sickness in Africa; (b) *T. cruzi*, which is the causative agent of Chagas disease in Latin America; and (c) a variety of *Leishmania* species that cause leishmaniasis in tropical and subtropical areas worldwide. These illnesses have been classified by the World Health Organization as neglected diseases, which affect people living in poverty in developing countries and for which no efficient therapy is available [17-19].

The Apicomplexa family encompasses a large group of protists, including approximately 5,000 known parasitic species, which are characterised by the presence of an apical complex containing a set of organelles involved in the infection process. Apicomplexan parasites infect invertebrate and vertebrate hosts, including humans and other mammals. The most serious parasitic disorder is caused by apicomplexan *Plasmodium* species, the etiological agent of malaria, which causes more than one million deaths annually [1]. Toxoplasmosis is another important disease caused by the apicomplexan parasite *Toxoplasma gondii*; it has been estimated that almost half of the human population worldwide is infected with this protozoa [20]. The life cycle of the apicomplexan parasites generally consists of complex asexual and sexual reproduction, but some differences are observable among distinct genera. Malaria transmission occurs during the blood feeding of the *Anopheles* mosquito, whereas toxoplasmosis is mainly transmitted by the ingestion of raw meat or contaminated cat feces.

Autophagy is a physiological self-degradative pathway essential for the maintenance of the metabolic balance in eukaryotes, leading to the turnover of cellular structures during both the normal cell cycle and during conditions of stress, such as starvation [21,22]. This process depends on double-membrane vesicles known as autophagosomes, which are responsible for the engulfment of macromolecules and organelles and the recycling of their components without an inflammatory response [23]. In eukaryotic cells, proteins known as Atgs contribute to the formation of autophagosomes and their targeting to lysosomes [24]. The autophagic machinery interfaces with many cellular pathways, such as that of the immune response and the inflammatory process, and acts as an inductor or suppressor of these processes [25]. Some molecules and organelles can undergo autophagy by specific proteins, such as in the selective pathway known as xenophagy, which is also observed in the degradation of intracellular pathogens [26,27]. The involvement of autophagy in this process has been demonstrated in the interactions of different pathogens with the host cells [28-30]. In protozoan infections, the role of autophagy has been debated in light of conflicting evidence presented in the literature, which tends to vary with the experimental model. Some studies suggest that parasites evade host cell defences using autophagy, while others suggest that the host uses autophagy to eliminate the pathogen [31-35]. However, there is no doubt that the autophagic machinery decisively influences the pathogenesis and virulence of protozoan infections; this machinery may therefore represent a promising target for drug discovery [36]. The autophagic process also occurs in the protozoa [37,38] and could occur in parallel to the host cell pathway, thus increasing the complexity of the phenomena. In the following sub-sections, the biology of

Trypanosomatidae and Apicomplexa protozoa will be reviewed in relation to the role of autophagy during the infection of the host cells.

2. Trypanosomatids and autophagy

As previously mentioned, the transmission of neglected diseases caused by trypanosomatids (sleeping sickness, Chagas disease and leishmaniasis) depends on an insect vector, and the environmental change from one host to another is a drastic event for the protozoa. To complete its life cycle, many metabolic and morphological changes must occur for the parasite to survive in a new host [39-42]. In addition to the kinetoplast, other characteristic ultrastructural structures are present in these parasites, including a single mitochondrion, unique flagella, sub-pellicular microtubules, glycosomes, acidocalcisomes and reservosomes (the last one is present exclusively in *T. cruzi*) [8]. In the context of the remodelling of sub-cellular structures, autophagy is greatly involved in eukaryotic homeostasis (including in that of trypanosomatids). However, the deregulation of this pathway, which is induced by conditions of stress, also leads to the parasite's death (Table 1). The sequencing of the complete genome of trypanosomatids has enabled the identification of parasitic genes [43-45]. Blast analysis comparing the trypanosome genome with yeast and mammalian genomes, with a particular emphasis on genes encoding autophagic machinery, has indicated the presence of some *ATG* genes in trypanosomatids [46,47]. However, the partial lack of a ubiquitin-like system, which is crucial for autophagosome formation, and the absence of cytoplasm-to-vacuole-targeting pathway orthologs suggest that these parasites have alternative autophagic features.

3. *T. brucei*

T. brucei is the etiological agent of sleeping sickness (or African trypanosomiasis) and is transmitted by the infected tsetse fly (*Glossina* sp.). After a blood feeding, procyclic trypomastigotes migrate from the insect midgut to the salivary gland where they undergo differentiation to infective metacyclic forms. Subsequently, these metacyclic trypomastigotes are inoculated into the mammalian host during the blood meal of the fly and differentiate into a proliferative bloodstream slender form. Interestingly, after a new differentiation, adapted short-stumpy forms evade the host immune system and disseminate the infection to the whole body; these forms are also able to cross the blood-brain barrier, which causes severe behavioural abnormalities, such as somnolence during daytime [48] (Figure 1). Unlike all other pathogenic trypanosomatids, which have an intracellular life-stage, *T. brucei* remains in the bloodstream of the mammalian host throughout the process of infection and, as such, is exposed to different environmental conditions that can trigger autophagy.

3.1. Role of autophagy in *T. brucei*

The first report on this parasite and autophagy was published in the 1970s by Vickerman and colleagues. These authors described the presence of myelin-like structures in different forms of the parasite observed by transmission electron microscopy [49, 50]. Many years later, it was suggested that the autophagic pathway is involved in the turnover of glycosomes during protozoan differentiation [51]. Glycosomes are peroxysome-like organelles that perform early glycolytic steps and are also involved in lipid metabolism. It was demonstrated that glycosome contents are altered depending on the form of the parasite, with many of these organelles being close to glysosomes during the differentiation process. A similar phenomenon was observed after nutrient deprivation of the parasite, reinforcing the fact that differentiation may cause the degradation of glycosomes by pexophagy.

Further genomic and bioinformatic analyses were performed that identified in *T. brucei* many *ATG* orthologs to those of yeasts and mammals [47,52]. These genes are involved in different steps of the autophagic pathway, such as induction (*ATG24*, *PEX14*, *TOR1* and *TOR2*, *VAC8*), vesicle nucleation (*ATG6*, *VPS15* and *VPS34*) and vesicle expansion and completion (*ATG3*, *ATG7*, *ATG9*, two isoforms of *ATG4* and *ATG8*). Two isoforms of Atg4 and two of Atg8 were recently characterised structurally [53], and it was postulated that Atg8.2 is essential for autophagosome formation and that Atg8 depletion is associated with delayed cell death [54].

It is thought that many drugs may trigger autophagy in African trypanosomes. Dihydroxya-cetone (DHA), spermine (snake venom) and vasoactive intestinal peptide (VIP – a neuropep-tide secreted by the immune system) induce the appearance of morphological features of autophagy in *T. brucei* [55-58]. DHA is an interesting compound to be used in therapy for sleeping sickness because its phosphorylation is DHA kinase-dependent, and DHA kinase is present in mammals and other eukaryotes but not in trypanosomes. After DHA uptake, this compound is not eliminated, leading to typical morphological characteristics of autophagy similar to those found in rapamycin treatment. In another report [59], the authors showed that hydrogen peroxide can produce the appearance of autophagic profiles, suggesting that the release of reactive oxygen species acts as a signal in the autophagic pathway in *T. brucei*, as it does in other eukaryotic cells [60-62].

4. *T. cruzi*

T. cruzi is the causative agent of Chagas disease. It is mainly transmitted by triatomine bugs, which are commonly known as "kissing bugs". In the insect midgut, proliferative forms of the parasite called epimastigotes differentiate to metacyclic trypomastigotes after migration to the posterior intestine. During the blood meal, triatomines eliminate urine and feces with infective trypomastigotes that then gain access to the vertebrate bloodstream. After internalisation in the host cell, trypomastigotes remain in parasitophorous vacuoles (PV) that fuse with lyso-somes, allowing an acidification of this compartment, which is an essential step towards differentiation into proliferative amastigotes. In the cytosol, successive parasite cycles occur until a new intracellular differentiation to trypomastigotes occurs; it is these forms that are responsible for the infection and dissemination to other cells and tissues [8] (Figure 2).

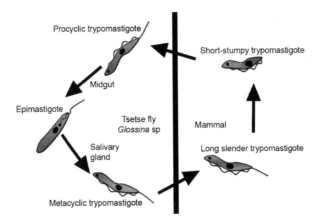

Figure 1. *T. brucei* life cycle.

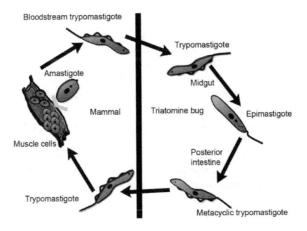

Figure 2. *T. cruzi* life cycle.

4.1. Role of autophagy in *T. cruzi*

Ultrastructural evidence of autophagy in *T. cruzi* was observed after the treatment of epimastigotes and bloodstream trypomastigotes with drugs; the appearance of myelin-like figures was the most recurrent feature detected [63-67]. Recently, the synergistic combination of amiodarone and posoconazole was able to trigger autophagy in replicative amastigotes [68].

In this way, different classes of therapeutic agents are able to induce the formation of auto-phagosomes, an event associated with parasite-related autophagic cell death, being the interplay between other programmed cell death as apoptosis or necrosis not discarded [69]. Due to the limitations of cell models, previous studies of different parasite forms have employed alternative techniques, such as monodansylcadaverine (MDC) staining and ATG gene expression, to demonstrate autophagy in the parasite [66,67]. Unfortunately, *T. cruzi* molecular machinery does not allow the use of double-stranded RNA to knock down target RNAs [70]; in addition, the lack of recognition of protozoan proteins by anti-Atg commercial antibodies hampers the evaluation of autophagy in this parasite. In spite of the advances in molecular and cellular biology, transmission electron microscopy remains a gold standard for autophagy analysis [71,72].

Aside from the description of autophagosomes in all *T. cruzi* life stages, description of the Atg cascade involved in autophagosome formation is not complete. Almost all *T. brucei ATG* genes have ortholog genes in *T. cruzi* [37,47]. In this parasite, two isoforms of Atg8 were described, with only Atg8.1 localised in autophagosomes as expected. These data suggest that there is only partially shared autophagic machinery, as is observed in human Atg8 orthologs [37]. In another study [37], the authors described the participation of *T. cruzi* Atg4 and Atg8 isoforms under conditions of nutritional stress and in the differentiation process from epimastigotes to metacyclic trypomastigotes, a process known as metacyclogenesis. The authors observed a remarkable expression of Atg8.1 by immunofluorescence microscopy, which was suggestive of intense autophagy in differentiating epimastigotes. Moreover, Atg8 co-localised with reservosomes, which are pre-lysosomal compartments related to energy supply that are present only in epimastigotes [73,74]. The reservosomal content consumed during metacyclo-genesis and the presence of Atg8 in this organelle strongly suggest that there is crosstalk between autophagy and reservosomes [75,76]. Transmission electron microscopy studies have produced images from endoplasmic reticulum profiles surrounding reservosomes that indicate the possible origin of preautophagosomal structures [66]. It is well known that PI3K inhibitors, such as 3-methyladenine and wortmannin, prevent autophagy in different experi-mental models [54,66]; however, these data are controversial due to a previous report dem-onstrating that treatment with kinase inhibitors staurosporine, genistein, 3-methyladenine and wortmannin led to the formation of autophagosomes [77]. The data indicate the necessity of careful use of PI3K inhibitors to block autophagy and the urgent need for the development of new specific autophagic inhibitors [78].

4.2. Host cell autophagy and *T. cruzi* infection

Though thought to be essential for parasite success, lysosomal fusion could be involved in autophagy during host cell interaction and might contribute to the process of degradation and elimination of *T. cruzi*. In 2009, the role of autophagy in parasite entry and co-localisation with the PV was described, resulting in increased infection of Chinese hamster ovary cells; this observation was subsequently confirmed in macrophage and heart cell lineages [34,79]. Starvation conditions and the addition of rapamycin led to an increase in the scale of the infection; this increase was partially reversed by 3-methyladenine, wortmannin and vinblastine, suggesting that autophagy favours the parasite during *T. cruzi*-host cell interactions. However, other groups demonstrated that classical autophagic stimuli (nutritional stress and rapamycin) did not produce an increase in parasite proliferation or even in the number of infected cells [33]. Recently, studies have emphasised role of autophagy in the control of *T. cruzi* infection using different cells and parasite strains (Figure 3) [80,81]. Once more, the conflicting data presented in the literature need to be further debated in light of the complexity of the protozoal strains and host cell models employed.

5. *Leishmania* species

The other medically important trypanosomatids are *Leishmania* species. Leishmaniasis is transmitted to mammals by sandflies, mainly of the *Phlebotomus* and *Lutzomia* genuses. Amastigotes differentiate into replicative procyclic promastigotes in the digestive tract of these sandflies, proliferate in the Phlebotominae gut, and then migrate to the proboscis where a new differentiation occurs to metacyclic promastigotes, the infective forms of the parasite. During the sandflies' blood meals, metacyclic promastigotes are inoculated into mammalian tissue and are phagocytised by macrophages. Inside the host cells, promastigotes differentiate into amastigotes that replicate and are responsible for cell lysis and dissemination in the organism (Figure 4). Currently, more than 20 species of *Leishmania* are known, each causing different clinical manifestations of the disease, including cutaneous leishmaniasis and visceral leishmaniasis (or Kala-azar). The pathogenicity depends on the *Leishmania* species and the host's immune response [8].

5.1. Role of autophagy in *Leishmania* sp.

Many groups have investigated autophagy cell death induced by drugs or antimicrobial peptides in various *Leishmania* species using electron microscopy and MDC staining [82-89]. Bioinformatics analysis has been a crucial checkpoint in the characterisation of *ATG* and *TOR* pathways in trypanosomatids [38,47,90]. In 2006, the role of autophagy in the differentiation process of *L. major* and *L. mexicana* was first evaluated [38,90]. The authors developed *L. major* VPS4, a mutant that could not complete the differentiation to the infective forms due to interference in autophagosome formation during conditions of starvation. The increase in Atg8 expression in differentiating forms supports the hypothesis that autophagy plays a pivotal role in metacyclogenesis [38,91]. In *L. mexicana*, the lack of cysteine peptidases CPA and CPB impairs autophagosomes formation and parasite differentiation; this finding is corroborated by the results of wortmaninn treatment and *ATG* deletion [90].

Recently, a subunit of protein kinase A in *L. donovani* that interferes with autophagy and protozoa differentiation was identified [92]. As observed in other trypanosomatids, the presence of Atg8-like proteins and their association with Atg4 in *Leishmania* species indicates that these proteins play a role in vesicle expansion [93]. Interestingly, the Atg5-Atg12 complex involved in autophagosome elongation was not previously detected [47], but recent studies have demonstrated its existence. It has also been shown that Atg5 deletion severally affects

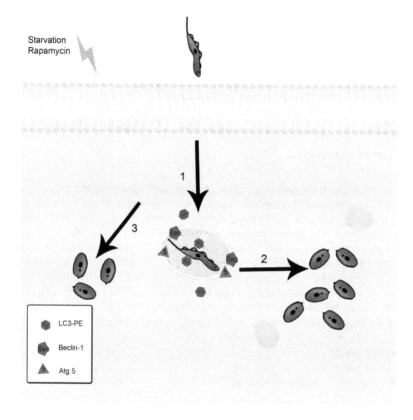

Figure 3. Autophagy in *T.cruzi*–host cell interaction. Romano et al [34] showed the co-localization of parasite vacuole with Atg proteins in the beginning of infection (1). Moreover, the replication of amastigotes is the same with or without autophagy induction (2) [33,34]. Rapamycin and starvation control infection reducing the number of amastigotes per cell (3) [80,81].

parasite homeostasis, producing a phenotype characterised by mitochondrial disruption, phospolipid accumulation and abnormal promastigote morphology [93,94]. Table 1 summarises the autophagic events in the three pathogenic trypanosomatids described in this chapter.

5.2. Host cell autophagy and *L. amazonensis* infection

The connection between the endosomal/lysosomal pathway and the PV results in macromolecules being taken up by the parasite, as demonstrated in *T. cruzi* infection [96]. In this context, a notable increase in the proliferation of *L. amazonensis* amastigotes was observed after autophagic induction by nutritional deprivation, rapamycin treatment or interferon-gamma. This mechanism was partially reversed by the autophagic inhibitors wortmaninn or 3-methyladenine, which significantly reduced amastigote replication (Figure 5) [33]. However,

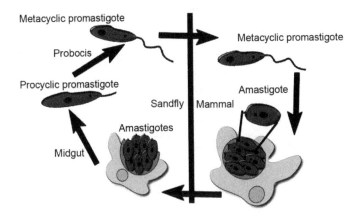

Figure 4. *Leishmania* sp. life cycle.

a recent report presented no correlation between the increase in LC3 expression and heightened *L. amazonensis* infection after treatment with autophagy inducers and inhibitors. In addition, macrophage autophagy was observed in inflammatory infiltrates of *L. amazonensis*-infected mice [97] and in natural human *L. donovani* infection [98].

6. Apicomplexa and autophagy

The phylum Apicomplexa comprises one of the most medically relevant groups of protists, which cause serious health and economic problems. Among these parasites, *Toxoplasma gondii* and *Plasmodium* species are well-known apicomplexans; it is estimated that malaria caused by *P. falciparum* kills over a million people annually. Another widespread disease is toxoplasmosis, which is caused by the apicomplexan parasite *T. gondii*; the severity of disease caused by this organism is directly related to patients' immunosuppression and is characterised by congenital transmission. In this context, knowledge of the detailed mechanisms involved in parasite infection and survival, including the role of autophagy, could contribute important information to the development of novel strategies for controlling Apicomplexa infections. Autophagy is an evolutionarily conserved pathway found in all eukaryotes, from unicellular organisms to metazoans; orthologs for approximately 30% of autophagy-related genes have been detected in apicomplexan sequenced genomes [99].

Among the key molecules involved in early autophagy steps, Atg1/ULK complex, Atg8 and Atg9 play crucial roles in cargo selectivity and in autophagosome formation [100,101]. Unlike other cell models, in Apicomplexa protozoa, the Atg8 C-terminal appears to not undergo processing before its association with phosphatidylethanolamine (PE) in the membrane of

Parasite	Life-stage	Phenotype	Stimuli	References
T. brucei	bloodstream trypomastigotes	autophagic cell death	DHA, neuropeptides, rapamycin, starvation	[55,58, 59]
	procyclic trypomastigotes	autophagic cell death	spermine (snake venom)	[57]
		Autophagy-induced differentiation	rapamycin, starvation	[54,56]
		unfolded protein response in endoplasmic reticulum associate with autophagy	DTT	[95]
T. cruzi	epimastigotes, trypomastigotes	autophagic cell death	SBIs; LPAs and cetoconazole; naphthoquinones; naphthoimidazoles; MBHA; posoconazole and amiodarone	[63-65,67, 68,71,72,]
	metacyclic trypomastigotes	Autophagy-induced differentiation	starvation; differentiation medium	[37]
L. amazonensis	promastigotes, amastigotes	autophagic cell death	amiodarone; elatol; lipophilic diamine	[83,86,89]
L. chagasi	promastigotes	autophagic cell death	yangambin	[87]
L. donovani	promastigotes	autophagic cell death	antimicrobial peptides; cryptolepine	[82,88]
L. major	promastigotes, amastigotes	autophagic cell death	cathepsin inhibitors	[85]
	metacyclic promastigotes	autophagy induces differentiation	differentiation medium; starvation	[38,91]
L. donovani	metacyclic promastigotes	autophagy induces differentiation	differentiation medium; starvation	[90]

Table 1. Summary of autophagic events in trypanosomatids. DHA: Dihydroxyacetone; DTT: dithiothretiol; SBIs: sterol biosynthesis inhibitors; LPAs: lysophospholipid analogues; MBHA: Morita–Baylis–Hillman adduct.

autophagosomes, suggesting a different regulation of this Atg protein in these organisms than in mammals and fungi [102]. Using a technique to detect lipidated Atg8 in *Plasmodium* species, only a single band corresponding to ATG8 was observed, suggesting that this parasite's Atg8 exists predominantly in the PE-conjugated form [22].

Two important kinases have opposing roles in the autophagic process: TOR (target of rapamycin) and class III phosphatidylinositol3-kinase (PI3K) [78,103]. In well-established autophagic models, TOR and class III PI3K represent negative and positive regulators, respectively, that act through complexes with regulatory subunits orchestrated by signalling cascades.

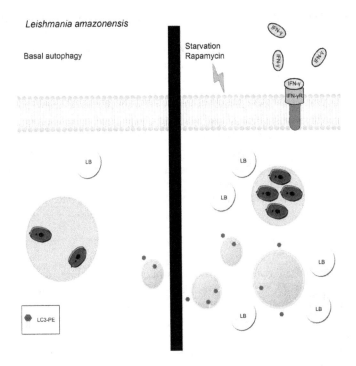

Figure 5. Autophagy in *L. amazonensis*-host cell interaction. When autophagy is induced, more amastigotes replicate and PV is smaller than in basal autophagic cells. Also, more lipid bodies are present, increasing infection and signaling to replication [33,97].

Analysis of the *T. gondii* genome revealed the presence of TOR and PI3K but not of other proteins crucial to the formation of these complexes [99]. Curiously, no genes for TOR complex machinery were found in the *Plasmodium* genome. Thus, it is possible that these unicellular eukaryotes have specific unknown proteins for several steps of the autophagic pathway instead of an absence of key proteins [22,104].

7. *T. gondii*

T. gondii is an obligate intracellular parasite with a complex life cycle involving one definitive feline host where the sexual phase occurs and intermediate hosts, such as birds, other mammals and man [105]. The main transmission routes to humans are the following: (i) the ingestion of raw meat containing tissue cysts (essentially bradyzoites forms); (ii) the ingestion of water and food contaminated with feline feces residue containing oocysts; and (iii) transplacentary pathway of tachyzoites [106]. After oral ingestion, tissue cysts or oocysts rupture, liberating the slow-replicating forms known as bradyzoites and sporozoites, respectively, which then invade intestinal epithelial cells. In the intracellular environment, the parasites differentiate into the fast-replicating tachyzoites that proliferate inside the host cell PV. The sustained infection depends on the modification of the PV membrane by the insertion of *T. gondii* secreted proteins, which prevent the fusion to lysosomes and, consequently, the elimination of the parasite (Figure 6) [20,107].

In healthy adults, *T. gondii* cysts are established in the host cells mainly in the eyes, brain and muscles during the chronic phase of toxoplasmosis [108]; however, in immunocompromised patients, such as HIV-positive patients, or in congenital toxoplasmosis, the disease becomes much more severe, and its complications could lead to death [20,109,110]. Despite the high percentage of people infected, the available therapy for toxoplasmosis is effective only in the tachyzoite stage and presents limited efficacy against the tissue cyst, which is the latent form of the parasite [111]. In this context, many efforts are necessary to develop new drugs to treat *T. gondii* infection [17].

7.1. Role of autophagy in *T. gondii* infection

Only a few studies on the *T. gondii* autophagic pathway have been performed, and these studies suggest opposing roles of autophagy in the parasite infection [102,112]. The presence of TgAtg8 in autophagic vesicles was observed in tachyzoites during their intracellular replication; similarly, severe parasite growth arrest due to TgAtg3 knockdown and recent identifications of the presence of TgAtg1 and TgAtg4 in the parasite suggest a role for autophagy in *T. gondii* homeostasis, although long-term exposure to autophagic stimuli was found to be harmful to the parasite (Figure 7) [112; 113].

Tachyzoites divide by a process called endodyogeny, whereby two daughter cells are developed inside a mother cell and leave residual material at the end of division. During this process, autophagy might be involved in recycling the mother cell organelles, such as micronemes and rhoptries, which are synthesised de novo in the daughter cells; however the accumulation of organelles after endodyogeny has not been observed in TgATG3 knockout organisms, making other experiments necessary to confirm this hypothesis [113]. One important phenotype detected in autophagic mutants is the loss of mitochondrial integrity [102,112]. Mitophagy, which is the autophagy of mitochondria, regulates the mitochondrial number to match metabolic demand; this process represents a quality control that is necessary for the removal of damaged organelles [114]. Autophagic stimuli are able to direct the mitochondrial network of tachyzoites towards their autophagic pathway, but the molecular machinery involved in selective targeting of the organelle remains unclear [102,112]. Nutrient deprivation has been shown to be a classic stimulus for the autophagic pathway activation in a large variety of organisms [37,115]. In *T. gondii* tachyzoites, starvation induces autophagy in extracellular and intracellular parasites [102,112]. Furthermore, autophagosomes were observed in parasites after a long extracellular nutritional restriction, suggesting that autophagy can act as a mechanism of resistance to starvation for nutrient recycling until the infection of a new host cell [102].

T. gondii life cycle

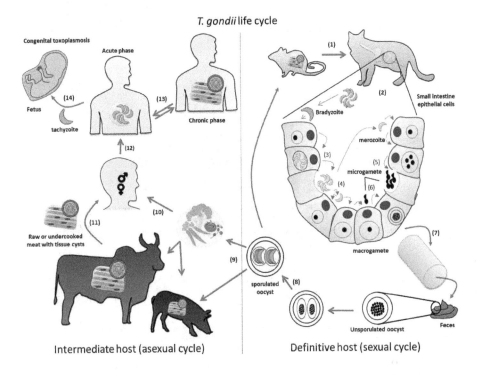

Intermediate host (asexual cycle)　　　Definitive host (sexual cycle)

Figure 6. *T. gondii* life cycle. (1) Definitive host infection; (2) Cyst disruption and intestinal epithelial cell infection; (3) Formation of merozoites; (4,5) Start of sexual phase with the formation of macrogametes and flagellate microgametes from merozoites; (6) Fusion of microgamete and macrogamete; (7) Oocyst release to the environment in the faeces; (8) The unsporulated oocysts become infective and contaminate the environment [116-118]; (9) The sporulated oocysts can cause infection of animals via consumption of contaminated food and water. (10,11) Human infection occurs by the ingestion of raw or undercooked meat of infected animals containing *T. gondii* cysts; (12) *T. gondii* tachyzoite multiplication in the intermediate host; (13) Tachyzoite-bradyzoite differentiation and formation of tissue cysts; (14) Transplacentary transmission of tachyzoites.

The data presented here demonstrate possible functions of *T. gondii* autophagy in parasite homeostasis. However, it has been proposed that, when strongly induced, the autophagic pathway represents a self-destructive mechanism leading to protozoal death. The molecular pathway of autophagic cell death is still unknown, and it is debated whether the pathway is a type of programmed cell death or a survival response to death stimuli [119]. Intracellular starved tachyzoites showed systematic mitochondrial fragmentation and a defect in host cell internalisation. As *T. gondii* is an obligate intracellular protozoa, the loss of invasion capacity leads to parasite death. The impairment in infective ability was related to the loss of mitochondrial integrity because organelles from apical complexes, such as rhoptries and micronemes, which are usually associated with the invasion process, are intact in these parasites [112]. Interestingly, these authors also demonstrated that autophagic inhibitor 3-methylade-

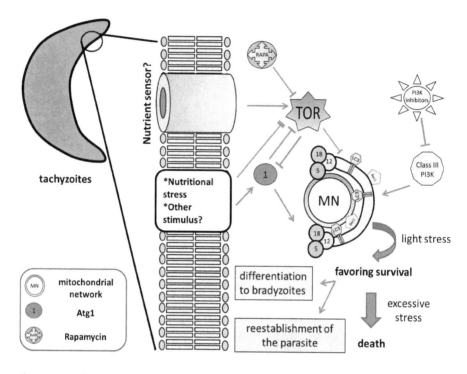

Figure 7. *T. gondii* tachyzoites response to autophagic stress. Autophagy acts in survival or death mechanisms in apicomplexan parasites depending on the environmental stress conditions. Arrows: activation; Headless arrows: inhibition.

nine prevented mitochondrial fragmentation, suggesting autophagic involvement in *T. gondii* death.

While nutritional stress has been extensively used as a model for autophagy, this condition is not easily encountered in the host cells and tissues *in vivo*. However, parasites could be exposed to nutritional restriction in the extracellular environment. The viability of tachyzoites kept in an axenic medium for periods of up to 12 hours drastically decreases, but a significant number of parasites nevertheless differentiate into bradyzoites [120]. Indeed, these observations raise the hypothesis that autophagy could be an adaptive mechanism of *T. gondii* to survive for short periods in starvation conditions, allowing the parasite to recover when favourable conditions occur or even to differentiate into a cystic form. Another interesting point for discussion is the correlation between mitochondrial fragmentation in intracellular tachyzoites and the depletion of amino acids in the culture medium [112]. Activated macrophages infected with the parasite showed low availability of the essential amino acid tryptophan, a condition that directly contributes to the protozoa's death in these cells [121,122]. In this context, TOR kinase is a vital component of the amino acid sensing mechanism in eukaryotic cells, as suggested by the detection of TgTOR by bioinformatic approaches and the evaluation of the activity of the classical TOR inhibitor rapamycin. This inhibitor triggered mitochondrial fragmentation of intracellular tachyzoites in starved parasites, and this phenotype was reversed by adding 3-methyladenine [112].

7.2. Host cell autophagy and *T. gondii* infection

As previously mentioned, *T. gondii* can infect any nucleated cell, but the parasite tropism principally involves nervous and muscular cells where the establishment of cystic forms is observed in chronic toxoplasmosis [111,123]. As was observed for *T. cruzi*-host cell interactions, controversial data on the importance of autophagy during *T. gondii* infection have been described in the literature; indeed, it has been suggested that autophagy can either control or facilitate parasite internalisation and proliferation [32,35,124-128]. Despite the relevance of muscular and nervous cells for the establishment of infection and for the course of the disease, very little has been reported on the role of autophagy in the progression of infection. As we will discuss in the next paragraphs, previous studies on the connection between the autophagic pathway and *T. gondii* infection were performed in macrophages, which are cells that play an important role in the immune response against this parasite [129].

Previous reports have shown that cellular immunity mediated by CD40 stimulation redirects the *T. gondii* to a lysosomal compartment via the autophagic route, resulting in the antimicrobial activity of the macrophage *in vitro* and *in vivo* [124,125]. *In vivo*, parasite elimination was dependent on GTPase p-47, IFN-γ, IGTP, and PI3K and culminated in the rupture of the parasite's membrane [125] (Figure 8). Additionally, the relationship between autophagy and the fusion of lysosomes with the *T. gondii* PV seems to be dependent on the synergy between TRAF6 signalling downstream of CD40 and TNF-α [126]. However, the IFN-γ/p47 GTPase-dependent elimination of the parasite by macrophages is independent of CD40/TNF signalling *in vitro*, demonstrating the primary role of IFN-γ in immunity against *T. gondii* in mice [127]. As observed in astrocytes, autophagy is activated to eliminate intracellular parasite debris and thus prevent the host cell death. Investigations in macrophages also indicated that the CD40-p21-Beclin 1 pathway is a CD40-dependent immunity route to mediating *in vivo* protection [128]. Similarly, Atg5 is required for damage to the PV membrane and removal of the parasite in primary macrophages stimulated by IFN-γ, despite the fact that no autophagosomes involving *T. gondii* have been detected. Atg5 also appeared crucial for *in vivo* p47 GTPase IIGP1 recruitment to the vacuole membrane induced by IFN-γ, suggesting an additional autophagy-independent role for Atg5 in the GTPase trafficking process [32]. In *T. gondii* infected astrocytes, the participation of autophagy has been shown to be indirect. The IFN-γ-stimulation of astrocytes infected with tachyzoites triggers the recruitment of p47 GTPases to the PV and usually leads to rupture of the vacuole and parasite membrane. In this case, autophagy acts by removing protozoal debris that accumulates in the cytoplasm and causes cell injury. Additionally, autophagy assists in antigen presentation through MHC class II in astrocytes, allowing an intracerebral immune response to parasite [130].

So far, little has been described regarding the involvement of autophagy in the interaction of *T. gondii* with nonprofessional phagocytes. In primary fibroblasts or Hela cells, infection with tachyzoites induced LC3 conjugation to PE, accumulation of LC3-containing vesicles close to the PV and an overexpression of beclin-1 and phosphatidylinositol-3-phosphate in the host cells in the mTOR-independent pathway. The infection of Atg5-deficient fibroblasts was reduced in physiological concentrations of amino acids, reinforcing the host cell autophagic role in the recovery of nutrients by the parasite. Because the classical function of autophagy involves recycling of various cellular components and because *T. gondii* depends on the uptake of many nutrients from the host cell, it has been proposed that the parasite may take advantage of the mammalian autophagic machinery to achieve successful infection [35]. Table 2 shows the host autophagic roles during *T. gondii* infection.

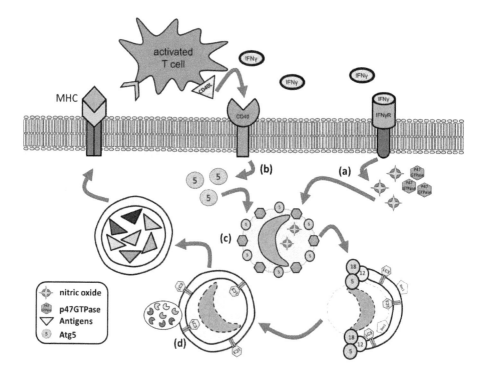

Figure 8. Autophagic role in *T. gondii* interaction with professional phagocytic cells. (a) INF-Y recruits P47GTPases to the PV membrane and induce nitric oxide production which limits the parasite replication. (b) CD40L activates Atg5 and recruits the autophagic machinery to the PV membrane. (c) PV and parasite membrane degradation by P47GTPase and Atg5. (d) Elimination of *T. gondii* debris by autophagolysosomal fusion and possible contribution of this process in antigen presentation through class II MHC.

Host cell	Induction	Phenotype	Reference
Peritoneal Macrophages and RAW264.7 lineage	CD40 stimulation and rapamycin	accumulation of LC3 around PV and low parasite load	[124]
Peritoneal macrophages	INF-γ stimulation	autophagy- dependent elimination of intracellular parasite debris	[125]
Peritoneal macrophages	INF-γ stimulation	Atg5-dependent PV membrane disruption	[32]
bone marrow Macrophages	CD40-p21-Beclin 1 pathway	stimulation of autophagy for protection against T. gondii	[128]
astrocytes	INF-γ stimulation	T. gondii debris removal by autophagy after vacuole and parasite membrane rupture by p47 GTPases	[130]
primary fibroblasts and Hela cells	T. gondii infection	Induction of LC3 conjugation to PE, accumulation of vesicles containing LC3 close to PV, beclin-1 and PI3K inside the cell	[35]

Table 2. Autophagy in *T. gondii*-host cell interactions

8. *Plasmodium* sp.

Plasmodium species are causative agents of malaria, the illness with the highest morbidity rate among human parasitic diseases. Currently, 5 species of *Plasmodium* sp. (*P. falciparum*, *P. vivax*, *P. malariae*, *P. ovale* and *P. knowlesi*) can infect humans, and lethality is associated with *P. falciparum* [131-133]. Sporozoites are transmitted by *Anopheles* sp. mosquitoes (definitive hosts) to the mammals (intermediate hosts), where they migrate primarily to the liver. After internalisation in hepatocytes, the parasites convert from elongated sporozoites (invasion competent and motile) to round proliferative trophozoites (metabolically active), which start the asexual reproduction process known as schizogony. At the end of the reproductive process, the daughter cells (merozoites) initiate maturation for erythrocyte invasion. When the merozoites become mature, they are enclosed in a membrane (the merosome) and released from hepatocytes to invade red blood cells, causing clinical symptoms of malaria (Figure 9). [135-137].

8.1. Role of autophagy in *Plasmodium* sp. infection

Recent publications have suggested that autophagy is involved in the differentiation of sporozoites to merosomes in hepatocytes [137,138]. The sporozoite-to-trophozoite differentiation is accompanied by the elimination of organelles unnecessary for schizogony and the production of merozoites in liver cells [137]. For example, micronemes and rhoptries are compartmentalised in the cytoplasm of sporozoites and sequestered in double-membrane structures resembling autophagosomes. In axenic conditions, the treatment of parasites with 3-methyladenine resulted in significant delay of the sporozoite differentiation process [139].

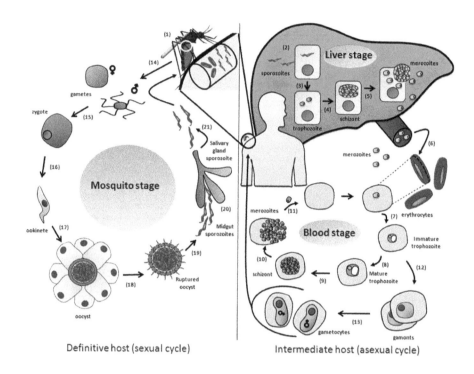

Figure 9. *Plasmodium* sp. life cycle. (1) Inoculation of sporozoites by malaria-infected female *Anopheles* mosquito into the human host. (2) Sporozoites infect hepatocytes. (3) Sporozoite-trophozoite differentiation. (4) Schizont formation. (5) Schizont rupture and release of merozoites. (6) Merozoites infect red blood cells. (7,8) Trophozoite maturation. (9) Schizont formation in red blood cells. (10) Schizont rupture and release of merozoites. (11) Infection of new red blood cells by the merozoites. (12,13) Differentiation of some parasites in gametocytes (sexual erythrocytic stages). (14) Ingestion of gametocytes by the mosquito during a blood meal. (15) Zygote formation in the mosquito´s stomach when the microgametes penetrate the macrogametes. (16) Zygote-ookinete differentiation. (17) Ookinetes invade the midgut wall of the mosquito where they develop into oocysts. (18,19) Oocysts rupture and release sporozoites. (20) Sporozoites migrate to the mosquito's salivary glands. (21) Mosquito inoculates sporozoites into a new human, perpetuating the parasite cycle.

After sporozoite differentiation, Atg8 is present in autophagosomes during the replication phase, suggesting an additional independent role for this protein in autophagy [137, 138,140].

The involvement of autophagy in *Plasmodium*-infected red blood cells has been poorly studied. One study demonstrated that erythrocytes infected with *P. falciparum* trophozoites and maintained in supplemented culture medium expressed Atg8 in the parasite cytosol. However, when these infected cells are submitted to restriction of glucose and amino acids, an increase in the number of autophagosomes labelled by Atg8 was observed, and these vesicles were found close to red blood cell membranes. Once erythrocytes no longer have organelles in the cytoplasm, the potential targets of autophagosomes in this cell model are debated. One hypothesis suggested that these autophagosomes target haemoglobin and blood nutrients to favour nutrient uptake by the parasite (Gaviria and colleagues, unpublished results). Surprisingly, no TOR ortholog was found in the *P. falciparum* genome (Sinai & Roepe, unpublished results), suggesting that under normal growth conditions, *P. falciparum* autophagy is configured as a survival process that is constitutively regulated by the acquisition of nutrients, although this route is unusual. Table 3 summarises the published reports on autophagic features in apicomplexans.

Little is known about the involvement of autophagy in the *Plasmodium sp.*-host cell interactions. So far, *Plasmodium ATG8* knock-out resulted in a lethal phenotype, indicating that this gene is essential for the mammalian life-stage of the parasite [22]. However, there have been no studies on the importance of the host cell autophagic machinery during the infection.

Parasite	Localisation	Induction	Phenotype	Reference
T. gondii	extracellular	Amino acid starvation	Basal: maintenance of life	[102]
	intracellular	Amino acid starvation and rapamycin	mitochondrial fragmentation	[111]
		Glucose and/or pyruvate starvation	Arrested mitochondrial fragmentation	
Plasmodium sp.	intracellular	sporozoite to trophozoite conversion in the liver	recycling of secretory organelles	[136]

Table 3. Autophagy in Apicomplexan parasites

9. Conclusion

The present chapter addresses the positive and negative regulations of the autophagic process of infected mammalian cells and the possible effects of these regulations on the *in vitro* and *in vivo* modulation of this process. This review also describes the autophagy pathway in pathogenic trypanosomatids and apicomplexans responsible for some of the most relevant neglected illnesses worldwide. The pivotal role of autophagy in pathogenicity and virulence was demonstrated in T. cruzi, T. brucei, Leishmania sp., T. gondii and Plasmodium sp., which suggests that autophagic machinery is a possible target for anti-parasitic intervention.

Acknowledgements

This work was supported with grants from CNPq (Universal), FAPERJ (APQ1) and IOC/FIOCRUZ.

Author details

Thabata Lopes Alberto Duque[1,2], Xênia Macedo Souto[1,3], Valter Viana de Andrade-Neto[4], Vítor Ennes-Vidal[5] and Rubem Figueiredo Sadok Menna-Barreto[1*]

*Address all correspondence to: rubemb@ioc.fiocruz.br

1 Laboratory of Cell Biology, Oswaldo Cruz Institute, Oswaldo Cruz Foundation, Rio de Janeiro, RJ, Brazil

2 Laboratory of Cell Biology, Department of Biology, Federal University of Juiz de Fora, MG, Brazil

3 Laboratory of Structural Biology, Oswaldo Cruz Institute, Oswaldo Cruz Foundation, Rio de Janeiro, RJ, Brazil

4 Laboratory of Biochemistry of Trypanosomatids, Oswaldo Cruz Institute, Oswaldo Cruz Foundation, Rio de Janeiro, RJ, Brazil

5 Laboratory of Molecular Biology and Endemic Diseases, Oswaldo Cruz Institute, Oswaldo Cruz Foundation, Rio de Janeiro, RJ, Brazil

Thabata Lopes Alberto Duque and Xênia Macedo Souto equally contributed to this work

References

[1] Nayyar GML, Breman JG, Newton PN, Herrington J. Poor-quality antimalarial drugs in southeast Asia and sub-Saharan Africa. Lancet Infectious Diseases 2012;12(6): 488-96.

[2] Soeiro MN, De Castro SL. *Trypanosoma cruzi* targets for new chemotherapeutic approaches. Expert Opinion on Therapeutic Targets 2009;13(1):105-21.

[3] Kobets T, Grekov I, Lipoldova M. Leishmaniasis: prevention, parasite detection and treatment. Current Medicinal Chemistry 2012;19(10): 1443-74.

[4] Welburn SC, Maudlin I. Priorities for the elimination of sleeping sickness. Advances in Parasitology 2012;79:299-337.

[5] Centers for Disease Control and Prevention. Toxoplasmosis. http://www.cdc.gov/parasites/toxoplasmosis/ (accessed 17 october 2012).

[6] World Health Organization. Working to overcome the global impact of neglected tropical diseases - First WHO report on neglected tropical diseases. Switzerland. 2010.

[7] Bañuls AL, Hide M, Prugnolle F. Leishmania and the leishmaniases: a parasite genetic update and advances in taxonomy, epidemiology and pathogenicity in humans.
Advances in Parasitology 2007;64:1-109.

[8] Stuart K, Brun R, Croft S, Fairlamb A, Gürtler RE, McKerrow J, et al. Kinetoplastids:
related protozoan pathogens, different diseases. The Journal of Clinical Investigation
2008;118(4): 1301-10.

[9] Greenwood BM, Fidock DA, Kyle DE, Kappe SH, Alonso PL, Collins FH, et al. Malaria: progress, perils, and prospects for eradication. Journal of Clinical Investigation
2008;118(4): 1266-76.

[10] Boyle JP, Radke JR. A history of studies that examine the interactions of *Toxoplasma*
with its host cell: Emphasis on in vitro models. International Journal of Parasitology
2009;39(8): 903-14.

[11] Teixeira AR, Gomes C, Lozzi SP, Hecht MM, Rosa AeC, Monteiro PS, et al. Environment, interactions between *Trypanosoma cruzi* and its host, and health. Cadernos de
Saúde Pública 2009;25 (1): S32-44.

[12] Nwaka S, Hudson A. Innovative lead discovery strategies for tropical diseases. Nature Reviews Drug Discovery 2006;5(11): 941-55.

[13] Hotez PJ, Bottazzi ME, Franco-Paredes C, Ault SK, Periago MR. The neglected tropical diseases of Latin America and the Caribbean: a review of disease burden and distribution and a roadmap for control and elimination. PLoS Neglected Tropical
Diseases 2008;2(9):e300.

[14] Le Pape P. Development of new antileishmanial drugs--current knowledge and future prospects. Journal of Enzyme Inhibition and Medicinal Chemistry 2008;23(5):
708-18.

[15] Nissapatorn V, Sawangjaroen N. Parasitic infections in HIV infected individuals: di-
agnostic & therapeutic challenges. The Indian Journal of Medical Research
2011;134(6): 878-97.

[16] Hotez PJ, Savioli L, Fenwick A. Neglected tropical diseases of the Middle East and
North Africa: review of their prevalence, distribution, and opportunities for control.
PLoS Neglected Tropical Diseases 2012;6(2):e1475.

[17] Lindoso JA, Lindoso AA. Neglected tropical diseases in Brazil. Revista do Instituto
de Medicina Tropical de São Paulo 2009;51(5): 247-53.

[18] Feasey N, Wansbrough-Jones M, Mabey DC, Solomon AW. Neglected tropical diseases. British Medical Bulletin 2010;93: 179-200.

[19] World Health Organization. http://www.who.int/neglected_diseases/en/ access in
November 8, 2012.

[20] Montoya JG, Liesenfeld, O. Toxoplasmosis. Lancet 2004;*363*(9425): 1965-76.

[21] Kiel JAKW. Autophagy in unicellular eukaryotes. Philosophical Transactions of the Royal Society 2010;365: 819-830.

[22] Brennand A, Gualdrón-López M, Coppens I, Rigden DJ, Ginger ML, Michels PAM. Autophagy in parasitic protists: Unique features and drug targets. Molecular and Biochemical Parasitology 2011;177 (2): 83-99.

[23] Levine B, Yuan J. Autophagy in cell death: an innocent convict? Clinical Investigation 2005;115(10):2679-88.

[24] Mitzushima N, Yoshimori T, Ohsumi Y. The role of Atg proteins in autophagosome formation. Annual Review of Cell Developmental Biology 2011; 27: 107-32.

[25] Levine B, Mizushima N, Virgin HW. Autophagy in immunity and inflammation. Nature 2011;469(7330): 323-35.

[26] Sumpter R, Levine B. Autophagy and innate immunity: triggering, targeting and tuning. Seminars in Cell & Developmental Biology 2010;21(7): 699-711.

[27] Kuballa P, Nolte WM, Castoreno AB, Xavier RJ. Autophagy and the immune system. Annual Review of Immunology 2012;30: 611-46.

[28] Gutierrez MG, Master SS, Singh SB, Taylor GA, Colombo MI, Deretic V. Autophagy is a defense mechanism inhibiting BCG and *Mycobacterium tuberculosis* survival in infected macrophages. Cell. 2004;119(6): 753-66.

[29] Deretic V, Levine B. Autophagy, immunity, and microbial adaptations. Cell Host & Microbe 2009;5(6): 527-49.

[30] Skendros P, Mitroulis I. Host cell autophagy in immune response to zoonotic infections. Clinical & Development Immunology 2012;2012: 91052. doi: 10.1155/2012/910525. (accessed 17 October 2012).

[31] Picazarri K, Nakada-Tsukui K, Nozaki T. Autophagy during proliferation and encystation in the protozoan parasite *Entamoeba invadens*. Infection & Immunity. 2008;76(1): 278-88.

[32] Zhao Z, Fux B, Goodwin M, Dunay IR, Strong D, Miller BC, et al. Autophagosome-independent essential function for the autophagy protein Atg5 in cellular immunity to intracellular pathogens. Cell host and Microbe 2008;4(5): 458-69.

[33] Pinheiro RO, Nunes MP, Pinheiro CS, D'Avila H, Bozza PT, Takiya CM, et al. Induction of autophagy correlates with increased parasite load of *Leishmania amazonensis* in BALB/c but not C57BL/6 macrophages. Microbes and Infection 2009;11(2): 181-90.

[34] Romano PS, Arboit MA, Vázquez CL, Colombo MI. The autophagic pathway is a key component in the lysosomal dependent entry of *Trypanosoma cruzi* into the host cell. Autophagy 2009;5(1): 6-18.

[35] Wang Y, Weiss LM, Orlofsky A. Host cell autophagy is induced by *Toxoplasma gondii* and contributes to parasite growth. The Journal of biological chemistry 2009;284(3): 1694-1701.

[36] Duszenko M, Ginger ML, Brennand A, Gualdrón-López M, Colombo MI, et al. Autophagy in protists. Autophagy. 2011;7(2): 127-58.

[37] Alvarez VE, Kosec G, Sant'Anna C, Turk V, Cazzulo JJ, Turk B. Autophagy is involved in nutritional stress response and differentiation in *Trypanosoma cruzi*. The Journal of Biological Chemistry. 2008; 283(6): 3454-64.

[38] Besteiro S, Williams RA, Morrison LS, Coombs GH, Mottram JC. Endosome sorting and autophagy are essential for differentiation and virulence of *Leishmania major*. The Journal of Biological Chemistry 2006;281(16): 11384-96.

[39] Saraiva EM, Pimenta PF, Brodin TN, Rowton E, Modi GB, Sacks DL. Changes in lipophosphoglycan and gene expression associated with the development of *Leishmania major* in *Phlebotomus papatasi*. Parasitology. 1995;111 (Pt 3): 275-87.

[40] Nolan DP, Rolin S, Rodriguez JR, Van Den Abbeele J, Pays E. Slender and stumpy bloodstream forms of *Trypanosoma brucei* display a differential response to extracellular acidic and proteolytic stress. European Journal of Biochemistry 2000;267(1): 18-27

[41] Gonçalves RL, Barreto RF, Polycarpo CR, Gadelha FR, Castro SL, Oliveira MF. A comparative assessment of mitochondrial function in epimastigotes and bloodstream trypomastigotes of *Trypanosoma cruzi*. Journal of Bioenergetics and Biomembranes 2011;43(6): 651-61.

[42] Castro DP, Moraes CS, Gonzalez MS, Ratcliffe NA, Azambuja P, Garcia ES. *Trypanosoma cruzi* immune response modulation decreases microbiota in *Rhodnius prolixus* gut and is crucial for parasite survival and development. PLoS One. 2012;7(5):e36591.

[43] Berriman M, Ghedin E, Hertz-Fowler C, Blandin G, Renauld H, Bartholomeu DC, et al. The genome of the African trypanosome *Trypanosoma brucei*. Science 2005;309(5733): 416-22.

[44] El-Sayed NM, Myler PJ, Bartholomeu DC, Nilsson D, Aggarwal G, Tran AN, et al. The genome sequence of *Trypanosoma cruzi*, etiologic agent of Chagas disease. Science 2005;309(5733): 409-15.

[45] Ivens AC, Peacock CS, Worthey EA, Murphy L, Aggarwal G, Berriman M, et al. The genome of the kinetoplastid parasite, *Leishmania major*. Science 2005;309(5733): 436-42.

[46] Rigden DJ, Herman M, Gillies S, Michels PA. Implications of a genomic search for autophagy-related genes in trypanosomatids. Biochemical Society Transactions 2005;33(Pt 5): 972-4.

[47] Herman M, Gillies S, Michels PA, Rigden DJ. Autophagy and related processes in trypanosomatids: insights from genomic and bioinformatic analyses. Autophagy 2006;2(2): 107-18.

[48] Hidron A, Vogenthaler N, Santos-Preciado JI, Rodriguez-Morales AJ, Franco-Paredes C, Rassi A. Cardiac involvement with parasitic infections. Clinical Microbiology Reviews 2010;23(2): 324-49.

[49] Brown RC, Evans DA, Vickerman K. Developmental changes in ultrastructure and physiology of *Trypanosoma brucei*. Transactions of Royal Society of Tropical Medicine Hygiene 1972;66(2): 336-7.

[50] Vickerman K, Tetley L. Recent ultrastructural studies on trypanosomes. Annales de la Société Belge de Médecine Tropicale 1977;57(4-5): 441-57.

[51] Herman M, Pérez-Morga D, Schtickzelle N, Michels PA. Turnover of glycosomes during life-cycle differentiation of *Trypanosoma brucei*. Autophagy 2008;4(3): 294-308.

[52] Barquilla A, Crespo JL, Navarro M. Rapamycin inhibits trypanosome cell growth by preventing TOR complex 2 formation. Proceedings of National Academy of Sciences U S A 2008;105(38): 14579-84.

[53] Koopmann R, Muhammad K, Perbandt M, Betzel C, Duszenko M. *Trypanosoma brucei* ATG8: structural insights into autophagic-like mechanisms in protozoa. Autophagy 2009;5(8): 1085-91.

[54] Li FJ, Shen Q, Wang C, Sun Y, Yuan AY, He CY. A role of autophagy in *Trypanosoma brucei* cell death. Cellular Microbiology 2012;14(8): 1242-56.

[55] Uzcátegui NL, Carmona-Gutiérrez D, Denninger V, Schoenfeld C, Lang F, Figarella K, et al. Antiproliferative effect of dihydroxyacetone on *Trypanosoma brucei* bloodstream forms: cell cycle progression, subcellular alterations, and cell death. Antimicrobial Agents and Chemotherapy 2007;51(11): 3960-8.

[56] Uzcátegui NL, Denninger V, Merkel P, Schoenfeld C, Figarella K, Duszenko M. Dihydroxyacetone induced autophagy in African trypanosomes. Autophagy 2007;3(6): 626-9.

[57] Merkel P, Beck A, Muhammad K, Ali SA, Schönfeld C, Voelter W, et al. Spermine isolated and identified as the major trypanocidal compound from the snake venom of *Eristocophis macmahoni* causes autophagy in *Trypanosoma brucei*. Toxicon 2007;50(4): 457-69.

[58] Delgado M, Anderson P, Garcia-Salcedo JA, Caro M, Gonzalez-Rey E. Neuropeptides kill African trypanosomes by targeting intracellular compartments and inducing autophagic-like cell death. Cell Death and Differetiation 2009;16(3): 406-16.

[59] Denninger V, Koopmann R, Muhammad K, Barth T, Bassarak B, Schönfeld C, et al. Kinetoplastida: model organisms for simple autophagic pathways? Methods in Enzymology 2008;451: 373-408.

[60] Zhang Y, Qi H, Taylor R, Xu W, Liu LF, Jin S. The role of autophagy in mitochondria maintenance: characterization of mitochondrial functions in autophagy-deficient S. cerevisiae strains. Autophagy 2007;3(4): 337-46.

[61] Chen Y, Gibson SB. Is mitochondrial generation of reactive oxygen species a trigger for autophagy? Autophagy 2008;4(2): 246-8.

[62] Scherz-Shouval R, Elazar Z. Regulation of autophagy by ROS: physiology and pathology. Trends in BiochemicalSciences 2011;36(1): 30-8.

[63] Braga MV, Magaraci F, Lorente SO, Gilbert I, de Souza W. Effects of inhibitors of Delta24(25)-sterol methyl transferase on the ultrastructure of epimastigotes of Trypanosoma cruzi. Microscopy and Microanalysis 2005;11(6): 506-15.

[64] Santa-Rita RM, Lira R, Barbosa HS, Urbina JA, de Castro SL. Anti-proliferative synergy of lysophospholipid analogues and ketoconazole against Trypanosoma cruzi (Kinetoplastida: Trypanosomatidae): cellular and ultrastructural analysis. The Journal of Antimicrobial Chemotherapy 2005;55(5): 780-4.

[65] Menna-Barreto RF, Corrêa JR, Pinto AV, Soares MJ, de Castro SL. Mitochondrial disruption and DNA fragmentation in Trypanosoma cruzi induced by naphthoimidazoles synthesized from beta-lapachone. Parasitology Research 2007;101(4): 895-905.

[66] Menna-Barreto RF, Corrêa JR, Cascabulho CM, Fernandes MC, Pinto AV, Soares MJ, et al. Naphthoimidazoles promote different death phenotypes in Trypanosoma cruzi. Parasitology 2009;136(5): 499-510.

[67] Fernandes MC, Da Silva EN, Pinto AV, De Castro SL, Menna-Barreto RF. A novel triazolic naphthofuranquinone induces autophagy in reservosomes and impairment of mitosis in Trypanosoma cruzi. Parasitology 2012;139(1): 26-36.

[68] Veiga-Santos P, Barrias ES, Santos JF, de Barros Moreira TL, de Carvalho TM, Urbina JA, et al. Effects of amiodarone and posaconazole on the growth and ultrastructure of Trypanosoma cruzi. International Journal of Antimicrobial Agents 2012;40(1): 61-71.

[69] Menna-Barreto RF, Salomão K, Dantas AP, Santa-Rita RM, Soares MJ, Barbosa HS, et al. Different cell death pathways induced by drugs in Trypanosoma cruzi: an ultrastructural study. Micron 2009;40(2): 157-68.

[70] DaRocha WD, Otsu K, Teixeira SM, Donelson JE. Tests of cytoplasmic RNA interference (RNAi) and construction of a tetracycline-inducible T7 promoter system in Trypanosoma cruzi. Molecular and Biochemical Parasitolollogy 2004;133(2): 175-86.

[71] Sandes JM, Borges AR, Junior CG, Silva FP, Carvalho GA, Rocha GB, et al. 3-Hydroxy-2-methylene-3-(4-nitrophenylpropanenitrile): A new highly active compound

against epimastigote and trypomastigote form of *Trypanosoma cruzi*. Bioorganic Chemistry 2010;38(5): 190-5.

[72] Benitez D, Pezaroglo H, Martínez V, Casanova G, Cabrera G, Galanti N, et al. Study of *Trypanosoma cruzi* epimastigote cell death by NMR-visible mobile lipid analysis. Parasitology 2012 139(4): 506-15.

[73] Soares MJ, Souto-Padrón T, De Souza W. Identification of a large pre-lysosomal compartment in the pathogenic protozoon Trypanosoma cruzi. Journal of Cell Science 1992;102 (Pt 1): 157-67.

[74] Soares MJ. The reservosome of *Trypanosoma cruzi* epimastigotes: an organelle of the endocytic pathway with a role on metacyclogenesis. Memórias do Instituto Oswaldo Cruz 1999;94 (1): 139-41.

[75] Figueiredo RC, Rosa DS, Soares MJ. Differentiation of *Trypanosoma cruzi* epimastigotes: metacyclogenesis and adhesion to substrate are triggered by nutritional stress. The Journal of Parasitology 2000;86(6): 1213-8.

[76] Alvarez VE, Kosec G, Sant Anna C, Turk V, Cazzulo JJ, Turk B. Blocking autophagy to prevent parasite differentiation: a possible new strategy for fighting parasitic infections? Autophagy 2008;4(3): 361-3

[77] Braga MV, de Souza W. Effects of protein kinase and phosphatidylinositol-3 kinase inhibitors on growth and ultrastructure of *Trypanosoma cruzi*. FEMS Microbiology Letters 2006;256(2): 209-16.

[78] Petiot A, Ogier-Denis E, Blommaart EF, Meijer AJ, Codogno P. Distinct classes of phosphatidylinositol 3'-kinases are involved in signaling pathways that control macroautophagy in HT-29 cells. The Journal of Biological Chemistry. 2000; 275(2): 992-8.

[79] Romano PS, Cueto JA, Casassa AF, Vanrell MC, Gottlieb RA, Colombo MI. Molecular and cellular mechanisms involved in the *Trypanosoma cruzi*/host cell interplay. IUBMB Life 2012;64(5): 387-96.

[80] Martins RM, Alves RM, Macedo S, Yoshida N. Starvation and rapamycin differentially regulate host cell lysosome exocytosis and invasion by *Trypanosoma cruzi* metacyclic forms. Cellular Microbiology 2011;13(7): 943-54.

[81] Maeda FY, Alves RM, Cortez C, Lima FM, Yoshida N. Characterization of the infective properties of a new genetic group of *Trypanosoma cruzi* associated with bats. Acta Tropica 2011;120(3): 231-7.

[82] Bera A, Singh S, Nagaraj R, Vaidya T. Induction of autophagic cell death in *Leishmania donovani* by antimicrobial peptides. Molecular and Biochemichal Parasitology 2003;127(1): 23-35.

[83] Dos Santos AO, Veiga-Santos P, Ueda-Nakamura T, Filho BP, Sudatti DB, Bianco EM, et al. Effect of elatol, isolated from red seaweed *Laurencia dendroidea*, on *Leishmania amazonensis*. Marine Drugs 2010;8(11): 2733-43.

[84] Santos AO, Santin AC, Yamaguchi MU, Cortez LE, Ueda-Nakamura T, Dias-Filho BP, et al. Antileishmanial activity of an essential oil from the leaves and flowers of *Achillea millefolium*. Annals of Tropical Medicine and Parasitology 2010;104(6): 475-83.

[85] Schurigt U, Schad C, Glowa C, Baum U, Thomale K, Schnitzer JK, et al. Aziridine-2,3-dicarboxylate-based cysteine cathepsin inhibitors induce cell death in Leishmania major associated with accumulation of debris in autophagy-related lysosome-like vacuoles. Antimicrobial Agents and Chemotherapy 2010;54(12): 5028-41.

[86] de Macedo-Silva ST, de Oliveira Silva TL, Urbina JA, de Souza W, Rodrigues JC. Antiproliferative, Ultrastructural, and Physiological Effects of Amiodarone on Promastigote and Amastigote Forms of *Leishmania amazonensis*. Molecular Biology International 2011; doi: 10.4061/2011/876021 (accessed 17 October 2012).

[87] Monte Neto RL, Sousa LM, Dias CS, Barbosa Filho JM, Oliveira MR, Figueiredo RC. Morphological and physiological changes in *Leishmania* promastigotes induced by yangambin, a lignan obtained from *Ocotea duckei*. Experimental Parasitology 2011;127(1):215-21.

[88] Sengupta S, Chowdhury S, Bosedasgupta S, Wright CW, Majumder HK. Cryptolepine-Induced Cell Death of *Leishmania donovani* Promastigotes Is Augmented by Inhibition of Autophagy. Molecular Biological International 2011;2011: 187850.

[89] Silva AL, Adade CM, Shoyama FM, Neto CP, Padrón TS, de Almeida MV, et al. In vitro leishmanicidal activity of N-dodecyl-1,2-ethanediamine. Biomedicine and Pharmacotherapy 2012;66(3): 180-6.

[90] Williams RA, Tetley L, Mottram JC, Coombs GH. Cysteine peptidases CPA and CPB are vital for autophagy and differentiation in *Leishmania mexicana*. Molecular Microbiology 2006;61(3): 655-74.

[91]] Besteiro S, Williams RA, Coombs GH, Mottram JC. Protein turnover and differentiation in *Leishmania*. International Journal for Parasitology 2007;37(10): 1063-75.

[92] Bhattacharya A, Biswas A, Das PK. Identification of a protein kinase A regulatory subunit from *Leishmania* having importance in metacyclogenesis through induction of autophagy. Molecular Microbiology 2012;83(3): 548-64.

[93] Williams RAM, Woods KL, Juliano L, Mottram JC, Coombs GH. Characterisation of unusual families of ATG8-like proteins and ATG12 in the protozoan parasite *Leishmania major*. Autophagy 2009;5(2): 159-172.

[94] Williams RAM, Smith TK, Cull B, Mottram JC, Coombs GH. ATG5 is Essential for ATG8-Dependent Autophagy and Mitochondrial Homeostasis in *Leishmania major*. Plos Pathogens 2012; 8 (5): 1-14.

[95] Goldshmidt H, Matas D, Kabi A, Carmi S, Hope R, Michaeli S. Persistent ER stress induces the spliced leader RNA silencing pathway (SLS), leading to programmed cell death in *Trypanosoma brucei*. PLoS Pathogens 2010;6(1):e1000731.

[96] Schaible UE, Schlesinger PH, Steinberg TH, Mangel WF, Kobayashi T, Russell DG. Parasitophorous vacuoles of *Leishmania mexicana* acquire macromolecules from the host cell cytosol via two independent routes. Journal of Cell Science 1999;112(Pt 5): 681-93.

[97] Cyrino LT, Araújo AP, Joazeiro PP, Vicente CP, Giorgio S. *In vivo* and *in vitro Leishmania amazonensis* infection induces autophagy in macrophages. Tissue & Cell 2012; doi: http://dx.doi.org/10.1016/j.tice.2012.08.003 (accessed 17 October 2012).

[98] Mitroulis I, Kourtzelis I, Papadopoulos VP, Mimidis K, Speletas M, Ritis K. *In vivo* induction of the autophagic machinery in human bone marrow cells during *Leishmania donovani* complex infection. Parasitology International 2009;58(4):475-7.

[99] Besteiro S. Which roles for autophagy in *Toxoplasma gondii* and related apicomplexan parasites? Molecular and Biochemical Parasitology 2012;184: 1-8.

[100] Xie Z, Nair U, Klionsky DJ. Atg8 controls phagophore expansion during autophagosome formation. Molecular biology of the cell 2008;19(8): 3290-8.

[101] Shvets E, Fass E, Scherz-Shouval R, Elazar Z. The N-terminus and Phe52 residue of LC3 recruit p62/SQSTM1 into autophagosomes. Journal of cell science 2008;121(Pt 16): 2685-95.

[102] Besteiro S, Brooks CF, Striepen B, Dubremetz JF. Autophagy protein Atg3 is essential for maintaining mitochondrial integrity and for normal intracellular development of *Toxoplasma gondii* tachyzoites. PLoS pathogens 2011;7(12): e1002416.

[103] Diaz-Troya S, Perez-Perez ME, Florencio FJ, Crespo JL. The role of TOR in autophagy regulation from yeast to plants and mammals. Autophagy 2008;4(7): 851-65.

[104] Rigden DJ, Michels PA, Ginger ML. Autophagy in protists: Examples of secondary loss, lineage-specific innovations, and the conundrum of remodeling a single mitochondrion. Autophagy 2009;5(6): 784-94.

[105] Dubey JP, Jones JL. *Toxoplasma gondii* infection in humans and animals in the United States. International journal for parasitology 2008;38(11): 1257-78.

[106] Furtado JM, Smith JR, Belfort R Jr., Gattey D, Winthrop KL. Toxoplasmosis: a global threat. Journal of global infectious diseases 2011;3(3): 281-4.

[107] Mordue DG, Hakansson S, Niesman I, Sibley, LD. *Toxoplasma gondii* resides in a vacuole that avoids fusion with host cell endocytic and exocytic vesicular trafficking pathways. Experimental parasitology 1999;92(2): 87-99.

[108] Sullivan WJ Jr., Jeffers V. Mechanisms of *Toxoplasma gondii* persistence and latency. FEMS microbiology reviews 2012;36(3): 717-33.

[109] Ambroise-Thomas P, Pelloux H. Toxoplasmosis - congenital and in immunocompromised patients: a parallel. Parasitology today 1993;9(2): 61-3.

[110] Mele A, Paterson PJ, Prentice HG., Leoni P, Kibbler CC. Toxoplasmosis in bone marrow transplantation: a report of two cases and systematic review of the literature. Bone marrow transplantation 2002;29(8): 691-8.

[111] Innes EA. A brief history and overview of *Toxoplasma gondii*. Zoonoses and public health 2010;57(1): 1-7.

[112] Ghosh D, Walton JL, Roepe PD, Sinai AP. Autophagy is a cell death mechanism in *Toxoplasma gondii*. Cellular microbiology 2012;14(4): 589-607.

[113] Besteiro S. Role of ATG3 in the parasite *Toxoplasma gondii*: Autophagy in an early branching eukaryote. Autophagy 2012;8(3): 435-7.

[114] Youle RJ, Narendra DP. Mechanisms of mitophagy. Nature reviews Molecular cell biology 2011;12(1): 9-14.

[115] Meijer AJ, Codogno P. Regulation and role of autophagy in mammalian cells. The international journal of biochemistry and cell biology 2004;36(12): 2445-62.

[116] Dubey JP, Frenkel JK. Cyst-induced toxoplasmosis in cats. The Journal of protozoology 1972;19(1): 155-7.

[117] Dubey JP, Frenkel JK. Experimental toxoplasma infection in mice with strains producing oocysts. The Journal of parasitology 1973; 59(3): 505-12.

[118] Dubey JP. Feline toxoplasmosis and coccidiosis: a survey of domiciled and stray cats. Journal of the American Veterinary Medical Association 1973; 162(10): 873-7.

[119] Denton D, Nicolson S, Kumar S. Cell death by autophagy: facts and apparent artefacts. Cell death and differentiation 2012;19(1): 87-95.

[120] Yahiaoui B, Dzierszinski F, Bernigaud A, Slomianny C, Camus D, Tomavo S. Isolation and characterization of a subtractive library enriched for developmentally regulated transcripts expressed during encystation of *Toxoplasma gondii*. Molecular and biochemical parasitology 1999;99(2): 223-35.

[121] Pfefferkorn ER. Interferon gamma blocks the growth of *Toxoplasma gondii* in human fibroblasts by inducing the host cells to degrade tryptophan. Proceedings of the National Academy of Sciences of the United States of America 1984;81(3): 908-12.

[122] Pfefferkorn ER, Eckel M, Rebhun S. Interferon-gamma suppresses the growth of *Toxoplasma gondii* in human fibroblasts through starvation for tryptophan. Molecular and biochemical parasitology 1986;20(3): 215-24.

[123] Black MW, Boothroyd JC. Lytic cycle of *Toxoplasma gondii*. Microbiology and molecular biology reviews 2000;64(3): 607-23.

[124] Andrade RM, Wessendarp M, Gubbels MJ, Striepen B, Subauste, CS. CD40 induces macrophage anti-*Toxoplasma gondii* activity by triggering autophagy-dependent fu-

sion of pathogen-containing vacuoles and lysosomes. The Journal of clinical investigation 2006;116(9): 2366-77.

[125] Ling Y M, Shaw MH, Ayala C, Coppens I, Taylor GA, Ferguson DJ, et al.. Vacuolar and plasma membrane stripping and autophagic elimination of *Toxoplasma gondii* in primed effector macrophages. The Journal of experimental medicine 2006;203(9): 2063-271.

[126] Subauste CS, Andrade RM, Wessendarp M. CD40-TRAF6 and autophagy-dependent anti-microbial activity in macrophages. Autophagy 2007;3(3): 245-8.

[127] Zhao Y, Wilson D, Matthews S, Yap GS. Rapid elimination of *Toxoplasma gondii* by gamma interferon-primed mouse macrophages is independent of CD40 signaling. Infection and immunity 2007;75(10): 4799-803.

[128] Portillo JA, Okenka G, Reed E, Subauste A, Van Grol J, Gentil K, et al. The CD40-autophagy pathway is needed for host protection despite IFN-Gamma-dependent immunity and CD40 induces autophagy via control of P21 levels. PLoS One 2010;5(12): e14472.

[129] Bogdan C, Rollinghoff M. How do protozoan parasites survive inside macrophages? Parasitology Today 1999;15(1): 22-8.

[130] Halonen SK. Role of autophagy in the host defense against *Toxoplasma gondii* in astrocytes. Autophagy 2009;5(2): 268-9.

[131] Cox-Singh J, Davis TM, Lee KS, Shamsul SS, Matusop A, Ratnam S, et al. *Plasmodium knowlesi* malaria in humans is widely distributed and potentially life threatening. Clinical infectious diseases 2008;46(2): 165-71.

[132] Marchand RP, Culleton R, Maeno Y, Quang NT, Nakazawa S. Co-infections of *Plasmodium knowlesi, P. falciparum,* and *P. vivax* among Humans and Anopheles dirus Mosquitoes, Southern Vietnam. Emerging infectious diseases 2011;17(7): 1232-9.

[133] William T, Menon J, Rajahram G, Chan L, Ma G, Donaldson S, et al. Severe *Plasmodium knowlesi* malaria in a tertiary care hospital, Sabah, Malaysia. Emerging infectious diseases 2011;17(7): 1248-55.

[134] Frevert U, Engelmann S, Zougbede S, Stange J, Ng B, Matuschewski K, et al. Intravital observation of *Plasmodium berghei* sporozoite infection of the liver. PLoS biology 2005;3(6): e192.

[135] Amino R, Thiberge S, Martin B, Celli S, Shorte S, Frischknecht F, et al. Quantitative imaging of *Plasmodium* transmission from mosquito to mammal. Nature Medicine 2006;12(2): 220-4.

[136] Sturm A, Amino R, van de Sand C, Regen T, Retzlaff S, Rennenberg A, et al. Manipulation of host hepatocytes by the malaria parasite for delivery into liver sinusoids. Science 2006;313(5791): 1287-90.

[137] Jayabalasingham B, Bano N, Coppens I. Metamorphosis of the malaria parasite in the liver is associated with organelle clearance. Cell Research 2010; 20: 1043-1059.

[138] Coppens I. Metamorphoses of malaria: the role of autophagy in parasite differentiation. Essays in biochemistry 2011;51: 127-36.

[139] Vaid A, Ranjan R, Smythe WA, Hoppe HC, Sharma P. PfPI3K, a phosphatidylinositol-3 kinase from *Plasmodium falciparum*, is exported to the host erythrocyte and is involved in hemoglobin trafficking. Blood 2010;115(12): 2500-7.

[140] Kitamura K, Kishi-Itakura C, Tsuboi T, Sato S, Kita K, Ohta N, Mizushima N. Autophagy-Related Atg8 Localizes to the Apicoplast of the Human Malaria Parasite *Plasmodium falciparum*. Plos one 2012;7(8): 1-10.

Chapter 2

Induction of Autophagy by Anthrax Lethal Toxin

Aiguo Wu, Yian Kim Tan and Hao A. Vu

Additional information is available at the end of the chapter

1. Introduction

Autophagy is an intracellular process whereby cells break down long-lived proteins and organelles and is morphologically characterized by the formation of many large autophagic vacuoles in cytoplasm [1]. This evolutionary process is conserved across all eukaryotic cells and is fundamentally important in normal and pathological cell physiology and development [2, 3]. Autophagy occurs constitutively at a basal level in quiescent cells but the process may be up-regulated during periods of starvation [4] and in response to other stress stimuli [5]. Many recent studies also suggest the increasing association of autophagy in numerous physiological and pathological conditions such as neurodegeneration, death of cancer cells, tissue formation and host cells response to pathogens [5].

The process of autophagy begins with the formation of isolation membrane or phagophore followed by sequestration of organelles or part of the cytoplasm to form autophagosome. The double-membrane autophagosome subsequently fuses with lysosome to form autolysosome where its content are degraded and released into the cytoplasm [6]. Several important autophagy related genes (*ATG*) that are critical for autophagosome formation have been identified recently. Microtubule-associated protein 1 light chain 3 (LC3) is the mammalian orthologue of yeast Atg8 that is required for autophagosome formation [7]. During autophagy, cytosolic form of LC3-I is processed into a lipidated LC3-II which is tightly associated with autophagosome membranes [8]. In addition, Atg8 has also been identified for its involvement in the expansion of isolation membrane [9]. The other protein complex that is essential for elongation of the isolation membrane is Atg5-Atg12 complex [10].

Vegetative *Bacillus anthracis* generates two essential virulence factors: the anthrax toxin and the poly-γ-D glutamic acid capsule [11]. The primary virulence factor is a secreted zinc-dependent metalloprotease toxin known as lethal factor (LF), which is introduced into the cytosol by protective antigen (PA) through its receptors on the cells [12]. LF exerts its toxic effect through the disruption of mitogen-activated protein kinase kinase (MAPKK) signalling pathway, which is essential in mounting an efficient and prompt immune response against the invading pathogen [13]. LF is also a potent inhibitor on many functions of immune cells such as macrophages, dendritic cells, neutrophils, T cells and B cells [14].

2. Anthrax lethal toxin induces autophagy

Our study has provided evidence that autophagy was involved in anthrax pathogenesis. These results are briefly described below.

2.1. Cells and induction of autophagy

RAW 264.7 murine macrophage cells were transfected with pEGFP-LC3. The transfected cells were treated with anthrax LT. The existence of autophagy was identified by immunoblotting, fluorescent punctuate counting, formation of acidic vacuoles, and viability of the LT treated cells.

2.2. Acridine orange staining showed increased Acidic Vacuoles (AVO)

Increased in AVO formation is a typical feature observed in cells undergoing autophagy [16]. Hence, we examined the effect of LT on AVO formation in RAW 264.7 cells by using lysosomotropic agent acridine orange (AO). RAW 264.7 cells treated with LT displayed a dose-dependent increase in AVO formation [17].

2.3. Increased GFP-LC3 punctuate when cells were treated with LT

Atg8 is an ubiquitin-like protein that undergoes conjugation process during autophagy and is determined to be an essential component for autophagy [7]. LC3 is the human orthologue of Atg8 and is also the most widely used protein marker for detecting autophagic organelles. During autophagy, cytosolic LC3-I is linked to phosphatidylethanolamine (PE) to form LC3-II and remain tightly bound to the autophagosomal membranes [8]. This process can be indirectly monitored through the use of reporter protein GFP conjugated to LC3 [18]. In order to determine if LT induces autophagy, we overexpressed GFP-LC3 in cells and observed for fluorescent punctuate distribution of GFP-LC3, which represent autophagosome formation.

Transfection of cells with GFP-LC3 for fluorescence microscopy analysis is widely used to detect autophagosome. Stable GFP-LC3 expressing RAW 264.7 cells were treated with anthrax LT for 2 hours and exhibited increased GFP-LC3 punctuates distribution whereas untreated cells displayed a diffuse GFP-LC3 appearance (Figure 1) [17]. These punctuate fluorescent dots indicate autophagosomes formation. Most of these fluorescent dots were probably autophagosomes as autolysosomes had weaker or no fluorescence signals due to the presence of lower LC3-II proteins [8]. The reduced LC3-II level in autolysosome may be possibly due to degradation or recycling back to cytosolic LC3-I [8]. The punctuate distribution of GFP-LC3 in LT treated cells were similar to those treated with rapamycin which serve as positive control for autophagy induction. Rapamycin binds to and inhibits mammalian target of rapamycin (mTOR), a negative regulator of autophagy [5]. Nutrients starvation is also able to trigger autophagy. Accordingly, cells incubated in nutrient free salt solutions, EBSS, for 2 hours showed punctuate distribution of GFP-LC3. Autophagy induced by nutrients starvation produced more intense fluorescence punctuates compared to rapamycin or LT treated cells [17].

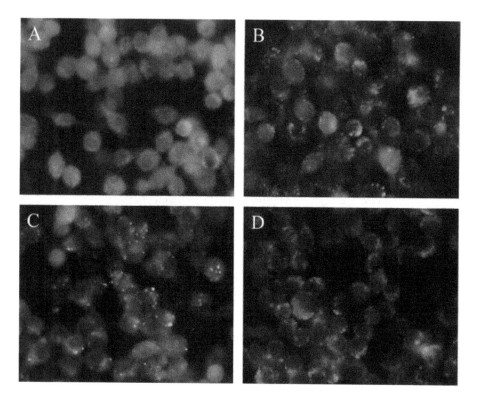

Figure 1. Lethal toxin induced punctuates EGFP-LC3 distribution in cells. Stably transfected RAW 264.7 cells expressing EGFP-LC3 were treated for 2 hours with (A) PBS, (B) 500ng/ml PA + 50ng/ml LF, (C) EBSS, (D) 4μM rapamycin. PBS, PA, LF and rapamycin were added directly into medium and EBSS treated cells were washed 3 times with PBS before incubation in EBSS. Images (40x) were taken from specimens under fluorescence microscope and are representative of 3 experiments.

2.4. Conversion of LC3-I to LC3-II

During autophagy, processing of cytosolic LC3-I to LC3-II permits autophagosomal membrane recruitment through an autophagic specific conjugation. As the amount of LC-II correlates with the extent of autophagosome formation [8], immunoblotting of LC3-II can be used to determine autophagy induction. To further corroborate that the GFP-LC3 punctuate observed was indeed autophagy induction by LT, we therefore examined the endogenous LC3-II levels in LT-treated cells.

RAW 264.7 cells were pre-treated with E64d and pepstatin A for 1 hour to inhibit lysosomal proteases followed by incubation with LT for 1 and 3 hours. At both time points, increase of LC3-II level from LT treated cells were detected, although the ratio differs from 1 to 3 hours after incubation (Figure 2A) [17]. It is not unusual to observe fluctuation of LC3-II level across various time point during autophagy induction [20].

We further determined if LT components could also induce autophagy individually. Treatment of cells with PA alone showed moderate increase of LC3-II while LF alone produced similar ratio of LC3-II / actin as control cells (Figure 2B) [17]. RAW 264.7 cells treated with LT for 1 hour showed elevated amount of endogenous LC3-II (Figure 2B). These observations are generally consistent with the fluorescent punctuate count in GFP-LC3 transfected RAW 264.7 cells. It may appear obvious that LF did not induce autophagy simply because it is not able to cross cell membrane in the absence of PA although it was reported that a small fragment of LF can enter into the cell cytoplasm without the assistance from PA [21]. Apparently, this mechanism of PA-independent insertion of LF into cytosol did not have an observable effect on autophagy induction at the concentration tested.

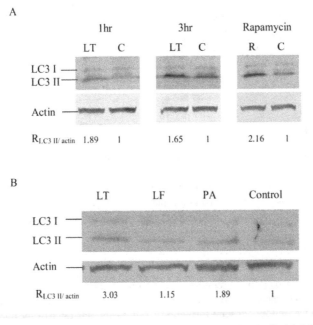

Figure 2. Immunoblot analysis of endogenous LC3-II conversion in RAW 264.7 cells. (A) Cells were pre-treated with 10µg/ml E64d and 10µg/ml Pepstatin A for 1 hour followed by incubation with LT (PA 500µg/ml + LF 100µg/ml) for 1 and 3 hours. Cells treated with 4µM rapamycin were used as positive control for autophagy induction. (B) Cells were pre-treated with E64d 10µg/ml and Pepstatin A 10µg/ml for 1 hour followed by incubation with 500ng/ml PA, 50ng/ml LF and LT (PA 500ng/ml + LF 50µg/ml) for another 1 hour. Total proteins were analysed by using anti-LC3 and anti-actin antibodies. Ratio of LC3 II/actin is shown under the blot.

2.5. Autophagy inhibitor may increase cell death

Inhibition of autophagy process can be used to investigate the role of autophagy in cellular response to toxins, bacteria or viruses. Depending on the interaction between autophagy mechanism and stimulus, the induction of autophagy may sometimes be beneficial or detrimental to the cells. Autophagy protects cells against *Vibrio cholera* cytolysin intoxication [15] but has an opposite effect when autophagy is activated in response to diphtheria toxin treatment [22]. Hence, we attempt to study the effect of autophagy on LT intoxication.

RAW 264.7 cells were treated with 10mM 3MA for 1 hour to inhibit autophagy followed by 2 and 3 hours of incubation with LT. Cells viability as determined by MTS assay showed no differences in 3MA treated and untreated cells (data not shown). This could be attributed to rapid lysis of RAW 264.7 cells when subject to LT treatment. Thus, we decided to use another cell line that is also susceptible to LT induced cell death but at a slower lysis rate than RAW 264.7. LT does not appear to cause instant lysis on human promyelocytic leukemia cell line HL-60 but is cytotoxic when HL-60 cells are differentiated into macrophage-like cells with PMA [23]. Differentiated HL-60 cells were pre-treated with 3MA for 1 hour followed by introduction of LT. Cells pre-treated with 3MA showed accelerated cell death compared to control cells at all the time points tested [17]. This suggests that autophagy may function as a defense mechanism against LT intoxication. While 3MA is often used as a specific inhibitor of autophagy [15], it also has effects on various aspects of metabolism that is unrelated to autophagy [24]. More studies need to be conducted to further understand the role of autophagy in LT intoxication.

3. Discussion

LT is recognized as a critical virulence factor in *B. anthracis* pathogenesis. Having been extensively researched for numerous years, LT pleiotropic actions on many cellular mechanisms have been described. Autophagy is activated during periods of physiological stress such as starvation as a means to sustain cell viability in a nutrient limiting environment [4]. In addition, autophagy is also implicated as a protective cellular response for the elimination of infectious agents [25]. However, certain pathogens are able to manipulate autophagy by altering certain processes for its survival and proliferation [25]. Recently, autophagy has become a rapidly growing biomedical marker as more studies unravel the role of autophagy in many physiological and pathological processes [6].

During autophagy, isolation membranes or phagophores elongate to sequester cytoplasmic components and become enclosed to form a double membrane autophagosome. Herein, we reported LT induced autophagosome formation in cells as demonstrated by the punctuate GFP-LC3 distribution in the cytoplasm and the corresponding increase in the punctuate counts. Another frequently used method as an indicator of autophagy is the monitoring of LC3-II conversion. LC3-II protein associates tightly to autophagosome and was determined to be correlated with autophagosome in cells [8]. Indeed, LT-treated cells displayed enhanced LC3-II conversion, which is a typical representative of autophagosome formation. As expected, PA was determined to be a critical component for autophagy induction. By itself, PA caused a moderate increase in LC3-II levels compared with non-treated controls. This could be attributed to a self-protection response of the host cells upon PA exposure. However, cells treated with LT (PA + LF) caused a dramatic increase in LC3-II levels [17]. This could be mainly the result of cellular stress and defence mechanism against the rapid toxic effects of LT. LT activity is believed to persist longer in the cells than PA alone, as indicated by its continuous enzymatic cleavage of substrate in the cells for 4-5 days [12, 26]. The prolonged presence of active LF in the cytoplasm may possibly play a contributing role in the dramatic increase of autophagy.

Autophagy may function as a defensive mechanism against toxins or invading pathogens but may also be exploited by microbes for survival/replication or even leading to death of host cells. In our study, autophagy was determined to be beneficial to differentiated human promyelocytic leukemia HL-60 cells exposed to LT as cells blocked from autophagy expressed accelerated cell death [17]. Probably similar to the cellular response to *V. cholerae* cytolysin intoxication [15], autophagy was presumably activated to enhance LT clearance from cytoplasm by diverting them to autophagosome and eventually eliminated by lysosomal degradation. As this study involved the use of cell lines, it is integral that the defensive role of autophagy be further determined on human primary macrophages or other immune cells. Other more specific autophagy gene knockdown/knockout studies can be carried out to confirm the results obtained from the commonly used autophagy inhibitor 3MA.

Meanwhile, circumstantial evidence from other non-autophagy related LT studies also suggests a possible link between lethal toxin and autophagy [27, 28]. As described earlier, autophagy proceeds from nascent vacuoles to become degradative autophagosomes by acquiring lysosomal proteins, including lysosome associated membrane protein (LAMP)-1 [29]. The maturation culminates with the subsequent fusion of the autophagome with lysosome to form autolysosome where it then degrades and releases its contents into the cytoplasm. LAMP-1 protein is also a major component of lysosome [30]. Kuhn et al analysed the proteomic profile of macrophages treated with LT and reported that LAMP1 protein was one of the highly upregulated protein [27], conceivably to increase lysosome capacity for fusing with autophagosomes and binding to late autophagosomes. In another separate study, several compounds were tested for its ability to modulate LT-induced cell death in macrophages [28]. Interestingly, the presence of rapamycin, an autophagy inducer, protected macrophages from LT-induced cell death. In contrast, macrophages co-treated with autophagy inhibitors, wortmannin or LY294002, exhibited accelerated cell death upon treatment with LT. Although autophagy was not part of their experimental design [28], it is worthy to note that the only compound tested in that study that protected macrophage from LT death in that experiment is a well known autophagy inducer, rapamycin. The results from these studies are in agreement with our current findings that LT activate autophagy and it may function as a cellular defense mechanism against LT intoxication.

Taken together, this study provides new insights into a hitherto undescribed effect of LT on cells; the induction of autophagic response in cells by PA and LT and the plausible role of autophagy in *B. anthracis* infection. Looking beyond, modulation of autophagy may potentially counter the detrimental effects of LT exposure in cells and remains a subject for further investigation.

Author details

Aiguo Wu[1], Yian Kim Tan[2] and Hao A. Vu[3]

1 Department of Molecular Microbiology, George Mason University, Manassas VA, USA

2 DSO National Laboratories, Singapore, Singapore

3 Biosecurity Research Institute, Kansas State University, Manhattan, USA

References

[1] R.A. Lockshin, Z. Zakeri, Apoptosis, autophagy, and more, Int. J. Biochem. Cell Biol. 2004;36(12) 2405-2419.

[2] B. Djehiche, J. Segalen, Y. Chambon, Inhibition of autophagy of fetal rabbit gono-ducts by puromycin, tunicamycin and chloroquin in organ culture, Tissue Cell 1996;28(1) 115-121.

[3] S. Tsukamoto, A. Kuma, M. Murakami, C. Kishi, A. Yamamoto, N. Mizushima, Au-tophagy is essential for preimplantation development of mouse embryos, Science 2008;321(5885) 117-120.

[4] W. Martinet, G.R. De Meyer, L. Andries, A.G. Herman, M.M. Kockx, In situ detection of starvation-induced autophagy, J. Histochem. Cytochem. 2006;54(1) 85-96.

[5] V. Deretic, Autophagy: an emerging immunological paradigm, J Immunol. 2012 189(1):15-20

[6] D.J. Klionsky, Autophagy: from phenomenology to molecular understanding in less than a decade, Nat Rev Mol Cell Biol 2007;8(11) 931-937.

[7] D.J. Klionsky, A.M. Cuervo, P.O. Seglen, Methods for monitoring autophagy from yeast to human, Autophagy 2007;3(3) 181-206.

[8] Y. Kabeya, N. Mizushima, T. Ueno, A. Yamamoto, T. Kirisako, T. Noda, E. Komina-mi, Y. Ohsumi, T. Yoshimori, LC3, a mammalian homologue of yeast Apg8p, is local-ized in autophagosome membranes after processing, EMBO J. 2000;19(21) 5720-5728.

[9] Z. Xie, U. Nair, D.J. Klionsky, Atg8 controls phagophore expansion during autopha-gosome formation, Mol. Biol. Cell 2008;19(8) 3290-3298.

[10] N. Mizushima, H. Sugita, T. Yoshimori, Y. Ohsumi, A new protein conjugation sys-tem in human. The counterpart of the yeast Apg12p conjugation system essential for autophagy, J. Biol. Chem. 1998;273(51) 33889-33892.

[11] M. Mock, A. Fouet, Anthrax, Annu. Rev. Microbiol. 2001;55 647-671.

[12] L. Abrami, S. Liu, P. Cosson, S.H. Leppla, F.G. van der Goot, Anthrax toxin triggers endocytosis of its receptor via a lipid raft-mediated clathrin-dependent process, J. Cell Biol. 2003;160(3) 321-328.

[13] J.F. Bodart, A. Chopra, X. Liang, N. Duesbery, Anthrax, MEK and cancer, Cell Cycle 2002;1(1) 10-15.

[14] D.J. Banks, S.C. Ward, K.A. Bradley, New insights into the functions of anthrax toxin, Expert Rev Mol Med 2006;8(7) 1-18.

[15] M.G. Gutierrez, H.A. Saka, I. Chinen, F.C. Zoppino, T. Yoshimori, J.L. Bocco, M.I. Colombo, Protective role of autophagy against Vibrio cholerae cytolysin, a pore-forming toxin from V. cholerae, Proc. Natl. Acad. Sci. U. S. A. 2007;104(6) 1829-1834.

[16] Y. Xu, S.O. Kim, Y. Li, J. Han, Autophagy contributes to caspase-independent macrophage cell death, J. Biol. Chem. 2006;281(28) 19179-19187.

[17] Y.K. Tan, C.M. Kusuma, L.J. St John , H.A. Vu HA, K. Alibek , A. Wu, Induction of autophagy by anthrax lethal toxin. Biochem Biophys Res Commun. 2009;379(2):293-7.

[18] N. Mizushima, Methods for monitoring autophagy, Int. J. Biochem. Cell Biol. 2004;36(12) 2491-2502.

[19] K.E. Beauregard, R.J. Collier, J.A. Swanson, Proteolytic activation of receptor-bound anthrax protective antigen on macrophages promotes its internalization, Cell Microbiol 2000; 2(3) 251-258.

[20] M.A. Delgado, R.A. Elmaoued, A.S. Davis, G. Kyei, V. Deretic, Toll-like receptors control autophagy, EMBO J. 2008;27(7) 1110-1121.

[21] N. Kushner, D. Zhang, N. Touzjian, M. Essex, J. Lieberman, Y. Lu, A fragment of anthrax lethal factor delivers proteins to the cytosol without requiring protective antigen, Proc. Natl. Acad. Sci. U. S. A. 2003;100(11) 6652-6657.

[22] K. Sandvig, B. van Deurs, Toxin-induced cell lysis: protection by 3-methyladenine and cycloheximide, Exp. Cell Res. 1992;200(2) 253-262.

[23] A. Kassam, S.D. Der, J. Mogridge, Differentiation of human monocytic cell lines confers susceptibility to Bacillus anthracis lethal toxin, Cell Microbiol 2005;7(2) 281-292.

[24] L.H. Caro, P.J. Plomp, E.J. Wolvetang, C. Kerkhof, A.J. Meijer, 3-Methyladenine, an inhibitor of autophagy, has multiple effects on metabolism, Eur. J. Biochem. 1988;175(2) 325-329.

[25] D. Schmid, C. Munz, Innate and adaptive immunity through autophagy, Immunity 2007;27(1) 11-21.

[26] S.D. Ha, D. Ng, J. Lamothe, M.A. Valvano, J. Han, S.O. Kim, Mitochondrial proteins Bnip3 and Bnip3L are involved in anthrax lethal toxin-induced macrophage cell death, J. Biol. Chem. 2007;282(36) 26275-26283.

[27] J.F. Kuhn, P. Hoerth, S.T. Hoehn, T. Preckel, K.B. Tomer, Proteomics study of anthrax lethal toxin-treated murine macrophages, Electrophoresis 2006;27(8) 1584-1597.

[28] M. Tsuneoka, T. Umata, H. Kimura, Y. Koda, M. Nakajima, K. Kosai, T. Takahashi, Y. Takahashi, A. Yamamoto, c-myc induces autophagy in rat 3Y1 fibroblast cells, Cell Struct. Funct. 2003;28(3) 195-204.

[29] W.A. Dunn, Jr., Studies on the mechanisms of autophagy: maturation of the autophagic vacuole, J. Cell Biol. 1990;110(6) 1935-1945.

[30] E.L. Eskelinen, Roles of LAMP-1 and LAMP-2 in lysosome biogenesis and autophagy, Mol. Aspects Med. 2006;27(5-6) 495-502.

Chapter 3

Infectious Agents and Autophagy: Sometimes You Win, Sometimes You Lose

Patricia Silvia Romano

Additional information is available at the end of the chapter

1. Introduction

Successful microorganisms are those that can evade the immune responses of the host. To reach this purpose, many pathogens have evolved as intracellular organisms, acquiring the capacity to live and to develop inside cells. This cellular parasitism has many benefits for pathogens such as protection from circulating antibodies, free access to nutrients and to specialized compartments that microorganisms use to establish their replicative niches. According to their particular lifestyles and requirements, many pathogens such as *L.monocytogenes*, *Shigella* or *T. cruzi*; reproduce in cell cytosol, while others target specific vesicles to generate particular compartments called parasitophorous vacuoles (PVs). This class of parasitism is utilized by *M. tuberculosis*, *C. burnetii* or *T. gondii*. On the other hand, in response to this level of adaptation, mammalian cells have developed different processes for eliminating intracellular microorganisms or for keeping them under strict control. These mechanisms are part of the innate immune responses, being phagocytosis (and the related processes) the best characterized. Innate cellular immunity also encompasses the autophagic process, a well conserved eukaryotic pathway that interacts with intracellular pathogens under certain circumstances to produce the destruction of the foreign organism. Autophagy comprises the inclusion of pathogens in autophagic-derived compartments and delivers them in lysosomes for digestion. Some pathogens, however, have acquired the capacity to subvert autophagy for their own benefit. This chapter will describe the interaction between intracellular microorganisms and the defense mechanisms of host cells, with special focus on the dual involvement of autophagy against pathogens, and the net outcome of this interaction.

2. The phagocytosis process

Phagocytosis is a form of endocytosis mainly present in specialized types of cells: the professional phagocytes that include the neutrophil, monocytes/macrophages, and dendritic cells. In the initial stage of phagocytosis, the cells change shape by sending out membrane projections (pseudopodia) that contact and surround the particle (bacteria, apoptotic bodies, etc) in a receptor-mediated and actin-dependent process (Figure 1). When the tips of the pseudopodia meet each other, membrane fusion occurs and the particle is enveloped in a vesicular compartment called phagosome. According to the type of internalized particle and the class of phagocyte involved, a different set of processes are activated that end in the destruction of the enclosed material. These mechanisms involve:

2.1. The respiratory or oxidative burst

The respiratory or oxidative burst is produced by the activity of specialized enzymes such as NADPH oxidase, which generates superoxide that recombines with other molecules as NO to form peroxynitrite, a potent oxidant agent against bacteria and parasites. Neutrophils and monocytes also utilize myeloperoxidase to further combine H_2O_2 with Cl^- to produce hypochlorite, a harmful component of phagosomes [1]

2.2. The production of microbicidal substances

Lysozime and defensins attack cell walls and membranes of certain bacteria.

2.3. Phagosome maturation

Phagosome maturation is a process that confers to nascent phagosomes the ability to kill pathogens or to degrade the ingested materials. Phagosomal maturation involves a complex sequence of reactions that result in the drastic remodeling of the phagosomal membrane and contents, produced as a consequence of vesicular fusion and fission events between the nascent vacuole and other cellular compartments mainly belonging to the endocytic pathway [2]. Rab and SNARES proteins are the main molecular components that regulate these events. Rabs are small GTP-binding proteins that control intracellular trafficking and supervise the maintenance of specific organellar identity, whereas SNARES are transmembrane proteins that, associating with their specific partners, form complexes that are the final executioners of the fusion processes between membranes. Interactions between phagosomes and endosomes commence soon after phagosome sealing, in a fashion that recapitulates the endocytic sequence: nascent (early) phagosomes (Eph) seemingly fuse initially with sorting (early) endosomes (EE), followed by late endosomes (LE) and ultimately lysosomes (Ly). Therefore, the membrane of Eph initially acquires components present in early endosomes such as Rab 5, phosphatidylinositol 3-phosphate (PI3P), Early Endosomal Antigen 1 (EEA-1) and Vamp-3. In contrast, late phagosomes (Lph) present Rab 7, Mannose-6-phosphate receptor, Vamp-7 and Lysosomal associated membrane proteins 1 and 2 (LAMPS 1 and 2) [2]. Furthermore, the luminal environment of phagosomes turned progressively more acidic due to the accumulation of H+ ATPase complexes in the phagosome membrane [3]. Phagolysosomes (PLy), the hybrid compartment generated by fusion between Lph and lysosomes, reach a pH of around 4.5, favoring the maturation of acidic hydrolases that will finally digest the materials (Figure 1, red compartments). Lysosomes contain several proteases, including Cathepsin D and Elastase, which are essential for killing various bacteria.

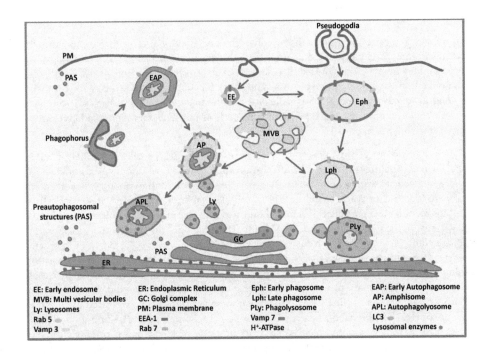

Figure 1. Phagocytosis and autophagy are the main cellular processes involved in the innate immune responses against pathogens. The scheme depicts the vesicular compartments that belong to each process and the main molecules that characterize them. Note that earlier, non-degradative compartments are colored in blue whereas acidic, lysosomal derived compartments are showed in red.

All these mechanisms generate a high level of protection against a wide range of pathogens. Paradoxically, phagocytosis can also have deleterious effects for the host: certain pathogens, exemplified by *Mycobacteria*, take advantage of the phagocytic machinery to gain access to the cell interior where, by subverting the maturation process, become intracellular pathogens [4–6].

Phagocytosis can also be produced in a class of cells different from immune cells. These "non-professional phagocytes" are cells with low phagocytic competence. Pathogens that can colonize these cells avoid the harmful ambient of the phagocyte-derived phagosome because many of the killing processes described for phagocytes are low or absent in non-professional phagocytes, keeping lysosomal degradation as the main defense system. In this way, the autophagic pathway which delivers cytoplasmic materials to lysosomes constitutes an important mechanism for eliminating pathogens, especially in these non-immune cells.

3. The autophagic pathway

Autophagy is a catabolic process that involves the degradation of cell components through the lysosomal machinery. Macroautophagy, the most studied type of autophagy, is important in many physiological situations such as development, cell growth, and cell differentiation. As a constitutive process, autophagy functions at basal levels in the turnover of long lived proteins and old organelles for maintaining cellular homeostasis. It can also be stimulated under different stress situations such as nutrient starvation, oxidative stress and intracellular infections [7].

The autophagic process involves specific compartments inside the cell. The initial preauto-phagosomal structures (PAS) are recruited to the cellular sites where autophagy is initiated [8]. A large number of studies have shown that specialized regions of the endoplasmic reticulum (ER) are involved in the formation of PAS [9,10]. However, more recent data indicate that besides ER other compartments like mitochondria, Golgi complex (GC) or plasma membrane (PM) may participate in this process [11–13]. The phagophorus, or isolation membrane, generated by fusion of PAS, is a curved membrane that in a way similar to that of the pseu-dopodia of macrophages, wraps the materials to be trapped which, in this case, consist of soluble or membranous content from cytoplasm (Figure 1, left). This membrane finally closes in a structure called autophagosome that transports the cargo for degradation. Immature (early) autophagosomes (EAP) are double-membrane vesicles easily recognized by electron microscopy [14]. Autophagosomes fuse with endocytic compartments (LE or multivesicular bodies, MVB) to form amphisomes (AP) that, in turn, fuse with lysosomes; forming autopha-golysosomes (APL) where the materials are degraded (Figure 1, red compartments).

At the molecular level, a large number of proteins engage in autophagy. The specific autophagy related proteins (Atgs) are a large family of proteins that regulate the nuclea-tion of PAS, and the formation and elongation of phagophorus and autophagosomes. At least 16 genes were found to be important for autophagy in yeasts, especially in the PAS nucleation [15]. Two protein conjugation reactions, both catalyzed by the action of Atg7, (an E3-like ubiquitin ligase activity), are mainly required for autophagosome formation in mammals. The mammalian Atg5-Atg12-Atg16L complex is recruited to the isolation membranes, favoring the elongation of the precursor membrane. The second conjugation system yields LC3-II which inserts into the autophagosomal membrane and contributes to vesicle elongation [16]. Pro-LC3 is initially cleaved by Atg4 to produce LC3-I. This molecule is a soluble protein distributed in the cytoplasm. After autophagic induction, LC3-I is conjugated with phosphatidylethanolamine (PE), allowing the insertion in the membrane of autophagic vesicles [16]. Two key signaling nodes converge to correlate autophagy with cell nutrient or stress conditions. The Tor-Atg1 signaling cascade transduces the response from growth factors, via class I PI3K, Akt/PKB, and so forth, to negatively regulate autophagy [17]. The second system, formed by Beclin1 (Atg6) and hVps34, is a lipid kinase that produces PI3P, which plays a pivotal role in early autophagosome formation, LC3 lipidation and the maturation of autophagosomes into autolysosomes [18].

Proteins that regulate transport and fusion events between vesicles are also important in autophagosome formation and maturation. Rab 7, a protein involved in transport to late endosomes and in the biogenesis of the perinuclear lysosome compartment is required for the normal progression of autophagosomes to autophagolysosomes [19]. The N-ethylmaleimide-sensitive factor attachment protein receptors (SNARES) Vamp3 and Vamp7 are important during the first steps of autophagy [20,21], whereas Vamp7 and Vamp8 also participate in the autophagosome-lysosome fusion [20,22]. Furthermore, it has been recently shown that actin has a role in the very early stages of autophagosome formation, linked to the PI3P generation step [23]. The description above shows that autophagosome formation and maturation engage similar molecular transport components and fusion machinery than that required for progression in endocytic and phagocytic pathways (Figure 1).

The process of autophagy can be monitored intracellularly by utilizing LC3 fused to a fluorescent protein (GFP-LC3 or mCherry-LC3). As the fluorescent LC3 is incorporated into autophagosomes, they can be seen as small puncta within the cell. As autophagy is a highly dynamic process, the number of puncta seen in a cell is a function of initiation as well as clearance (lysosomal fusion and subsequent degradation) [24]. Autophagosome initiation can be inhibited by blocking Class III PI3 kinases (Vps34) with 3-methyladenine or wortmannin, or by knockdown of essential factors such as Atg5 or Beclin-1, a component of Vps34 kinase complex. Autophagosome clearance can be prevented by interfering with lysosomal fusion by Bafilomycin A1, chloroquine, and other agents that tend to alkalinize the lysosome (e.g. NH_4Cl). Rapamycin inhibits the TOR signaling pathway, leading to induction of autophagy. Spermidine and resveratrol have been recently characterized as autophagy inducers. Genetic and functional studies indicate that spermidine inhibits histone acetylases, while resveratrol activates the histone deacetylase Sirtuin 1. Although it remains elusive whether the same histones (or perhaps other nuclear or cytoplasmic proteins) act as the downstream targets of spermidine and resveratrol, these results point to an essential role of protein hypoacetylation in autophagy control [25]. This hipoacetylated protein status leads to upregulation of several atg genes, including atg7, atg11 and atg15 in several organisms such as mammals, yeasts, nematodes and flies [26].

As explained above, host autophagy is a component of the innate responses against intracellular pathogens that generally functions as a second barrier when phagocytic or other defense mechanisms are exceeded. However, some pathogens have the capacity to evade autophagic responses or to subvert the autophagic pathway and to live and replicate inside autophagosomal compartments. The following sections describe the opposite effects of autophagic response against microorganisms.

3.1. Autophagy as a component of the innate immune responses

In order to internalize host cells, many pathogens induce their own ingestion in phagocytic cells. After entry, pathogenic microorganisms manipulate the normal (or "canonical") phagocytic pathway to evade lysosomal degradation and to achieve the maximal benefits (protection, nutrition, survival, and replication) from cells. These actions include inhibition of phagosome maturation, escape from phagosome to cytoplasm and development in a vacuole with particular characteristics. As a component of immune responses, autophagy hampers these mechanisms enclosing viruses, bacteria or parasites in compartments which share characteristics and molecular machinery with canonical autophagosomes, a process usually named as xenophagy [27]. For a better comprehension of the autophagic action, each pathogenic strategy will be described, and the most characteristic pathogen of each group will be exemplified in the following paragraphs.

- *The Mycobacteria case*: A marquee feature of the powerful human pathogen *Mycobacterium tuberculosis* is its macrophage parasitism. The intracellular survival of this microorganism rests upon its ability to arrest phagolysosome biogenesis, avoid direct cidal mechanisms in macrophages, and block efficient antigen processing and presentation. Lipoarabinomannan (LAM) and phoshatidylinositol mannoside (PIM) are two toxins elaborated by *Mycobacterium tuberculosis* that stimulate fusion between phagosomes and early endosomes and prevent Rab conversion on phagosomes by interference with Rab effectors, especially the type III PI3K (hVps34). LAM abolishes the normal recruitment of PI3K to mycobacterium phagosomal membrane decreasing the levels of PI3P. The final result is the reduction of the recruitment of EEA-1 and other effectors and the inhibition of the normal progression from early to late phagosomes [28]. Thus, a critical feature of the *M. tuberculosis* phagosome is its lack of the vacuolar H+ ATPase [29] and mature lysosomal hydrolases, such as Cathepsin D. In stark contrast, the induction of autophagy by physiological, pharmacological or immunological signals, including the major antituberculosis Th1 cytokine IFN-gamma, can overcome mycobacterial phagosome maturation block. Almost ten years ago, Gutierrez and colleagues demonstrated that when infected macrophages were treated in conditions that induce autophagy, mycobacterium-containing phagosomes become more acidic and also acquire markers of maturation, including the vacuolar H+ ATPase, LAMP-1, LBPA and Cathepsin D. Additionally, starvation promotes recruitment to mycobacterial phagosomes of critical autophagy components such as LC3, indicating that these phagosomes are redirected to a compartment with autophagic characteristics that finally fuses with lysosomes [30]. The most remarkable finding of this work was the demonstration that autophagy induction hampered the survival of this intracellular pathogen, recognizing to autophagy as an effector of innate immunity (see Figure 2A).

- *The group A Streptococcus case:* The second case belongs to pathogens that escape from phagosomes and turn into cytosolic invaders. The Group A of *Streptococcus* (GAS) is often internalized into nonphagocytic epithelial cells via the endocytic pathway. At early times after infection, GAS secrets its major virulence factor: the cytolysin streptolysin O (SLO) that supports the escape of GAS into the cytoplasm from endosomes [31]. After escaping, GAS became enveloped by autophagosome-like compartments and were killed upon fusion of these compartments with lysosomes. In contrast, in autophagy-deficient Atg5-/- cells, GAS survived, multiplied, and were released from the cells [32,33]. Additionally a SLO-deficient mutant of GAS was viable for a longer time than the wild-type strain, although it failed to escape the endosomes [31]. Both results highlight the crucial role of autophagy in the suppression of intracellular survival of this pathogen. A similar conclusion was recently obtained with *Staphylococcus aureus*. After invasion of non-phagocytic cells, virulent strains of this gram positive bacterium stimulate autophagy and become entrapped in intracellular PI3P-enriched vesicles and its effector WIPI-1, a protein present in the membrane of both phagophores and autophagosomes. This interaction seems to be deleterious for bacteria, given that these autophagosome-like WIPI-1 positive vesicles that envelope *S. aureus* are finally targeted for lysosomal degradation [34].

- *The Toxoplasma case*: The third strategy used by pathogenic organisms is to create a specialized compartment that remains isolated from the host endocytic or phagocytic networks. *Toxoplasma gondii* relies on this mechanism; the membrane of its PV is nonfusogenic due to its unique composition lacking host proteins. Nonetheless, macrophages infected with *Toxoplasma* can reroute the pathogen-containing compartment to lysosomes. Autophagy

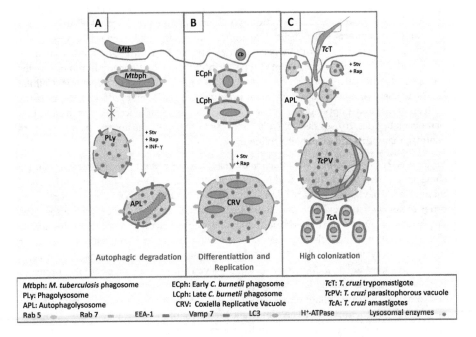

Figure 2. Mechanisms of pathogen-host autophagy interactions. As a component of innate immune responses against intracellular pathogens, autophagy can effectively eliminate some pathogens, re-routing them to lysosomes (autophagolysosomes). That is the case of *Mycobacterium tuberculosis* (A). In contrast, some microorganisms such as *Coxiella burnetii* utilize autophagic compartments to delay lysosomal fusion until differentiating into more resistant forms (B). On the other hand, *Trypanosoma cruzi* exploits the autophagic pathway to efficiently colonize host cells (C).

plays a key role in this process [35,36]. When CD40 of human or mouse macrophages infected with *T. gondii* is stimulated with (CD154+) CD4+ T cells or exposed to anti-CD4 antibodies, the nonfusogenic nature of the PV is reversed and the vacuole fuses with late endosomes and lysosomes. This fusion is dependent on autophagy, as indicated by the inhibition of this mechanism in cells knockdown for Beclin 1 or treated with 3-MA, an inhibitor of phagophore formation. CD40 activation also stimulates expression of LC3 that localizes to the PV [36].

3.2. Evasion of autophagic responses

Despite the potent effect of autophagy in killing intracellular pathogens, some microbial pathogens have the capacity to control cellular autophagy and successfully parasitize eukaryotic cells. These highly evolved microorganisms have developed specific virulence factors to protect themselves from autophagic elimination by producing:

• *Prevention of autophagy induction:* Several viruses direct their products to essential autophagic proteins, causing them to be functionally inhibited. The herpes virus family can produce autophagy blockage through different mechanisms. The HSV-1 ICP34.5 viral protein encoded by herpes simplex virus type 1 blocks Beclin1 function and confers neurovirulence in mice [37]. A similar mechanism was recently shown for human cytomegalovirus; the virus protein TRS1 interacts with Beclin 1 to inhibit autophagy [38]. Gamma-herpesviruses, including important human pathogens such as Epstein Barr virus or Kaposi's sarcoma-associated HIV, displayed a different type of inhibition. They encode homologs of the antiapoptotic, host Bcl-2 protein to promote viral replication and pathogenesis. Cellular Bcl-2 and their viral homologs have the property to bind and inhibit Beclin1, suppressing both apoptotic and autophagic responses [39]. It is not yet clear whether other intracellular pathogens besides viruses also actively suppress initiation of the autophagy pathway. In contrast, many bacteria display the following actions.

• *Suppression of autophagosome maturation into autolysosome:* Similar to mycobacterium phagosome maturation arrest, other pathogens have the ability to suppress autophagosome maturation. They specifically reside in vacuoles with autophagosomal characteristics in order to survive and replicate, but avoid transient or permanent fusion with lysosomes. *Porphyromonas gingivalis*, a bacterial periodontal pathogen that can localize to atherosclerotic plaques, traffics to autophagosomes as a way of evading the conventional endocytic trafficking to lysosomes [40], After intracellular uptake, *P. gingivalis* transits from early autophagosomes to late autophagosomes and prevents the formation of autolysosomes, a mechanism not yet well elucidated [41]. On the other hand, a delay in the delivery of lysosomal enzymes to phagosomes was initially described for dimorphic bacteria and named "the pregnant pause". The dimorphic life cycles of these pathogens have dramatic consequences for phagosome traffic. In the transmissible state, *C. burnetii, L. pneumophila* and others, such as *Leishmania sp.*, block phagosome maturation; after a pregnant pause that includes the bacterial differentiation process, replicative forms emerge and thrive in lysosomes [42]. Autophagy is one of the mechanisms activated by these intracellular pathogens for delaying lysosomal fusion. At late stages of cellular infection, both *Coxiella burnetii* and *Legionella pneumophila* develop vacuoles that have characteristics of phagolysosomes and are also decorated with LC3 [43,44]. In the case of *C. burnetii*, acquisition of LC3 is even an early event in the transit of phagosomes containing bacterium (Cph) and depends on bacterial protein synthesis because chloramphenicol avoid this LC3 recruitment [45]. Interactions with autophagic and also late endocytic compartments are maintained during the transit of Cph and in the development of the Coxiella replicative vacuoles (CRV) [46]. Indeed, autophagy induction or the overexpression of autophagic proteins LC3 and Rab 24 favor the generation and maturation of this CRV [47]. Taking together, these results demonstrated that *C. burnetii* transits through the normal endo/phagocytic pathway but actively interacts with autophagosomes at early times after infection. This intersection delays fusion with the lysosomal compartment, possibly favoring the intracellular differentiation and survival of the bacteria. In this period, *C. burnetii* differentiates from the transmissible forms (named small cell variant) to the replicative forms (large cell variant) (Figure 2B). In the case of *L. pneumophila*, it was recently demonstrated that this bacterium produced several Type IV effector proteins that control the timing of bacteria during intracellular transport. The early secretion of DrrA/SidM, LidA, and RalF factors, prolong association with the ER and permit the persistence of the bacteria in immature autophagosomal vacuoles for a period sufficient to differentiate into an acid-resistant, replicative form. Subsequent secretion of LepB releases the block of autophagosome maturation, and the adapted progeny continue to replicate within autophagolysosomes [48].

• *Evasion of pathogen recognition by the autophagic machinery:* This strategy is especially important in intracytoplasmic pathogens such as *Shigella flexneri*, *L. monocytogenes*, and *Burkholderia pseudomallei*. Shigella VirG, a protein required for intracellular actin-based motility, induced autophagy and favored the microorganism trap by autophagosomes, after binding between VirG and Atg5. However, *Shigella*, encoding Type III secretion effector, IcsB, competitively binds to Atg5, thereby camouflaging its own bacterial target molecule VirG from autophagic capture [49]. Furthermore, BopA, the counterpart of IcsB in Burkholderia pseudomallei, have similar autophagy-evading properties [50]. *Listeria monocytogenes* is a classic example of a "cytosol-adapted pathogen"; it can rapidly escape from the phagosome of macrophages and other non-phagocytic cells and replicate rapidly in the cytosol. Phagosome escape also enables cell-to-cell spread by the bacteria through a bacterial driven actin-based motility mechanism. Besides Act A, that as was shown plays a critical role in autophagic escape by polymerizing actin which favors bacteria movements [51], another virulence factor of *L. monocytogenes*, InlK, was recently shown to counteract the autophagic process. InlK interacts with the Major Vault Protein (MVP), the main component of cytoplasmic ribonucleoproteic particules named vaults. The recruitment of MVP to bacterial surface disguises intracytosolic bacteria from the autophagic recognition system leading to an increased survival rate [52].

3.3. Autophagy as a survival mechanism

A different type of host-microbial interaction belongs to the group of organisms that harness cell autophagy. Independently of the final localization within the cell, these particular organisms improve their intracellular cycle when interacting with autophagic compartments. Postulated benefits of host autophagy for microbes include the promotion of viral replication or morphogenesis via utilization of the autophagic machinery. Polyovirus localized in double-membrane autophagosome-like structures positive for LC3 serve as lipid membrane-scaffolds that enhance viral replication [53]. In a similar way, the rotavirus NSP4 protein colocalizes with LC3-positive vesicular compartments and is postulated to play a role in the formation of viroplasms and/or the packaging or transcription of the rotavirus genome [54].

Another mechanism is the utilization of autophagosomes as a protective intracellular niche to enhance the survival and growth of bacteria. As described above, dimorphic bacteria such as *C. burnetii* or *L. pneumophila* follow this method. In contrast, *Francisella tularensis*, enters LC3-positive compartments to allow cytoplasmic bacteria to regain access to the endocytic compartment to finally promote bacteria egress through exocytosis [55]. Autophagy could also favor intracellular pathogen survival by providing nutrients to pathogens, particularly those that reside in sequestered vacuoles that lack access to cytoplasmic nutrients. That is the case of *T. gondii*, which establishes its vacuole in the vicinity of autophagic compartments and that displays an impaired growth in *Atg5*-deficient MEF cells, leading to the conclusion that host cell autophagy plays a role in promoting parasite growth through nutrient recovery [56].

A special type of pathogen-autophagy interaction is produced by the protozoan parasite *Trypanosoma cruzi*. This pathogen exploits the autophagic pathway to efficiently colonize host cells, as will be described in detail in the next section.

3.4. Autophagy as an invasion strategy: The *Trypanosoma cruzi* case

The protozoan parasite *Trypanosoma cruzi* can invade a wide range of phagocytic and non-phagocytic cells in the infective, non-proliferative trypomastigote form. Inside the cell, trypomastigotes are temporarily contained in a membrane vesicle, the parasitophorous vacuole. Subsequently, the parasites escape to the cytosol, differentiate into the amastigote form, and replicate by binary division [57,58]. This replication process culminates after 9 cycles, followed by a new differentiation period where the parasites undergo a transition back into trypomastigotes. After that the parasites are released from the cell and infect the neighboring cells, maintaining the infection process.

The characteristics of the *T. cruzi* parasitophorous vacuole (TcPV) are directly related to the parasite invasion mechanism. Previously published data showed that two main invasion processes involving different signaling pathways are participating: the calcium dependent fusion of lysosomes with the host plasma membrane [59,60], and the activation of class I PI3K that produces a plasma membrane-derived vacuole initially devoid of lysosomal markers [61]. Although both pathways require the disruption of the host cell actin cytoskeleton, the lysosomal independent *T. cruzi* entry model appears to be more significant early after internalization (50% *versus* 20% lysosome-dependent entry process). However, lysosomal fusion is essential for the establishment of a productive infection [62] and for the progression and completion of the *T. cruzi* intracellular cycle [63,64], since the parasite Tc-Tox is activated in the acid environment provided by the lysosomes [57,65]. These strategies of cell invasion indicate that *T. cruzi* entry is a complex process that employs different components from the plasma membrane and the endocytic pathway to finally produce the intracellular infection. More recently, studies provided by our laboratory demonstrated that the *T. cruzi* parasitophorous vacuole (TcPV) is decorated with LC3 protein and that the autophagic inhibitors wortmannin, 3-methyladenine or vinblastine suppress this recruitment and also significantly reduce the intracellular infection. In contrast, induction of autophagy before infection by starvation or other means significantly increased the percentage of infected cells. Interestingly, infection was diminished in the absence of the specific autophagy genes Beclin1 or Atg5, which are required for initiation of autophagy, indicating that autophagic-derived compartments are required for efficient entry of *T. cruzi* into the host cell [66]. Live imaging using confocal microscopy showed that GFP-LC3 positive vesicles move towards plasma membrane and contact the sites where trypomastigotes, the *T. cruzi* invasive forms, bind the membrane [66,67] (Figure 2C). The pro-pathogen effects of autophagy on *T. cruzi* infection were observed in different classes of cells and *T. cruzi* strains, demonstrating that this interaction is a wide-spread phenomenon [68]. The autophagy modulation of host cells during the following stages of the *T. cruzi* intracellular cycle -trypomastigote to amastigote differentiation, amastigote replication and amastigote differentiation back to trypomastigote- seems to suffer no mdification compared to cells maintained in control conditions [66]. Since *T. cruzi* is an unicellular eukaryotic organism that also has its own autophagic pathway [69], other experimental procedures will be necessary to decipher the possible dual action of autophagy modulation on *T. cruzi* infected cells. Indeed, unpublished data from our laboratory show that classical inducers and inhibitors of mammalian autophagy have similar effects on *T. cruzi* and that protozoan autophagy is activated during *T. cruzi* metacyclogenesis, a process that renders metacyclic trypomastigotes from epimastigotes and that takes place in the digestive apparatus of the triatomine vectors.

3.5. Autophagy in action: *in vivo* infections studies

To date the outcome of the pathogen/autophagy relationship on *in vivo* infections models, with the exception of a few cases, remains little understood. Actions of autophagy as an innate immune component are easier to understand, particularly with the use of knockout mice. In this way, studies on *in vivo* M. *tuberculosis* murine infections showed that the most susceptible mice are those deficient in either IFN-γ or IFN-γ receptors [70]. These results clearly demonstrate that macrophage activation, and the macrophage autophagic pathway [30], are required as a critical components for controlling infection.

The main concerns with the *in vivo* models arise from the cases of pathogens that *in vitro* studies show to be favored by autophagy induction. At the moment, no current evidence demonstrates that autophagy gene deletion in the host attenuates microbial disease. Therefore, the physiological significance of microbial utilization of autophagy for "promicrobial" effects remains to be established [71]. The discrepant conclusions between *in vitro* and *in vivo* studies in *T. gondii* infection models exemplify this concept. Although *T. gondii* has impaired growth in *Atg5*-deficient cells, leading to the conclusion that host cell autophagy plays a role in promoting parasite growth through nutrient recovery [56], this parasite has increased virulence in mice with macrophage-specific deletion of Atg5 [72]. In agreement with these results, unpublished data from our laboratory show that autophagy-impaired mice are more susceptible to *T. cruzi* infection, while our previously results on cell cultures clearly demonstrate that decreased autophagic levels in Atg5 KO cells or in Beclin 1 KD cells significantly decreased *T. cruzi* infection [66].

One possible explanation for these discrepancies is the different effects of autophagy (or autophagic proteins) on phagocytic and non-phagocytic cells. When *T. gondii* infected macrophages were stimulated with CD40 receptor agonists, the parasite vacuole fuses with endosomes and lysosomes in an autophagy-dependent process leading to parasite destruction [35,36]. In contrast, in the non-phagocytic HeLa cells, Wang and colleagues reported the beneficial effects of autophagic induction for parasite survival and growth [56]. Considering this possibility, the comparative analysis of autophagic modulation on the course of a specific pathogen infection in phagocytic and non-phagocytic cells *in vitro* prior to mice infection would be productive in the future. However, this simplistic point of view will never replace the conclusions obtained from mice experiments, especially when immune responses are implicated.

Acknowledgements

I want to especially thank Dr. María Isabel Colombo, who introduced me in the world of pathogens and also shared with me her great knowledge about cell biology and autophagy. Our laboratory was supported by grants from Universidad Juan A. Maza, Comisión Nacional Salud Investiga (Ministerio de Salud de la Nación), CONICET-Fundación Fulbright, and CONICET-Fundación Bunge y Born. Our current grants are provided by Secretaría de Ciencia, Técnica y Posgrado (Sectyp, Universidad Nacional de Cuyo), CONICET (PIP 2011-2013) and Agencia Nacional de Promoción Científica y Tecnológica (PICT# 2008 2235) to Patricia S. Romano.

Author details

Patricia Silvia Romano*

Address all correspondence to: promano@fcm.uncu.edu.ar

Laboratory of Trypanosoma cruzi and the Host Cell, Institute of Histology and Embryology, National Council of Scientific and Technical Research (IHEM-CONICET), School of Medicine-National University of Cuyo. Mendoza, Argentina

References

[1] Winterbourn, C. C, & Kettle, A. J. (2012). Redox Reactions and Microbial Killing in the Neutrophil Phagosome. Antioxid Redox Signal

[2] Vieira, O. V, Botelho, R. J, & Grinstein, S. (2002). Phagosome maturation: aging gracefully. Biochem J , 366, 689-704.

[3] Hackam, D. J, Rotstein, O. D, Zhang, W. J, Demaurex, N, Woodside, M, Tsai, O, & Grinstein, S. (1997). Regulation of phagosomal acidification. Differential targeting of Na+/H+ exchangers, Na+/K+-ATPases, and vacuolar-type H+-atpases. J Biol Chem , 272, 29810-29820.

[4] Duclos, S, Diez, R, Garin, J, Papadopoulou, B, Descoteaux, A, Stenmark, H, & Desjardins, M. (2000). Rab5 regulates the kiss and run fusion between phagosomes and endosomes and the acquisition of phagosome leishmanicidal properties in RAW 264.7 macrophages. J Cell Sci 113 Pt , 19, 3531-3541.

[5] Hackstadt, T. (2000). Redirection of host vesicle trafficking pathways by intracellular parasites. Traffic , 1, 93-99.

[6] Meresse, S, Steele-mortimer, O, Moreno, E, Desjardins, M, Finlay, B, & Gorvel, J. P. (1999). Controlling the maturation of pathogen-containing vacuoles: a matter of life and death. Nat Cell Biol 1: EE188., 183.

[7] Meijer, A. J, & Codogno, P. (2004). Regulation and role of autophagy in mammalian cells. Int J Biochem Cell Biol , 36, 2445-2462.

[8] Noda, T, Suzuki, K, & Ohsumi, Y. (2002). Yeast autophagosomes: de novo formation of a membrane structure. Trends Cell Biol , 12, 231-235.

[9] Axe, E. L, Walker, S. A, Manifava, M, Chandra, P, Roderick, H. L, Habermann, A, Griffiths, G, & Ktistakis, N. T. (2008). Autophagosome formation from membrane compartments enriched in phosphatidylinositol 3-phosphate and dynamically connected to the endoplasmic reticulum. J Cell Biol , 182, 685-701.

[10] Matsunaga, K, Morita, E, Saitoh, T, Akira, S, Ktistakis, N. T, Izumi, T, Noda, T, & Yoshimori, T. (2010). Autophagy requires endoplasmic reticulum targeting of the PI3-kinase complex via Atg14L. J Cell Biol , 190, 511-521.

[11] Hailey, D. W, Rambold, A. S, Satpute-krishnan, P, Mitra, K, Sougrat, R, Kim, P. K, & Lippincott-schwartz, J. (2010). Mitochondria supply membranes for autophagosome biogenesis during starvation. Cell , 141, 656-667.

[12] Moreau, K, Ravikumar, B, Puri, C, & Rubinsztein, D. C. (2012). Arf6 promotes autophagosome formation via effects on phosphatidylinositol 4,5-bisphosphate and phospholipase D. J Cell Biol , 196, 483-496.

[13] Yamamoto, H, Kakuta, S, Watanabe, T. M, Kitamura, A, Sekito, T, Kondo-kakuta, C, Ichikawa, R, Kinjo, M, & Ohsumi, Y. (2012). Atg9 vesicles are an important mem-

brane source during early steps of autophagosome formation. J Cell Biol , 198, 219-233.

[14] Baba, M, Takeshige, K, Baba, N, & Ohsumi, Y. (1994). Ultrastructural analysis of the autophagic process in yeast: detection of autophagosomes and their characterization. J Cell Biol , 124, 903-913.

[15] Suzuki, K, Kubota, Y, Sekito, T, & Ohsumi, Y. (2007). Hierarchy of Atg proteins in pre-autophagosomal structure organization. Genes Cells , 12, 209-218.

[16] Kabeya, Y, Mizushima, N, Ueno, T, Yamamoto, A, Kirisako, T, Noda, T, Kominami, E, Ohsumi, Y, & Yoshimori, T. (2000). LC3, a mammalian homologue of yeast Apg8p, is localized in autophagosome membranes after processing. EMBO J , 19, 5720-5728.

[17] Cheong, H, Yorimitsu, T, Reggiori, F, Legakis, J. E, Wang, C. W, & Klionsky, D. J. (2005). Atg17 regulates the magnitude of the autophagic response. Mol Biol Cell , 16, 3438-3453.

[18] Fujita, N, Itoh, T, Omori, H, Fukuda, M, Noda, T, & Yoshimori, T. (2008). The Atg16L complex specifies the site of LC3 lipidation for membrane biogenesis in autophagy. Mol Biol Cell , 19, 2092-2100.

[19] Gutierrez, M. G, Munafo, D. B, Beron, W, & Colombo, M. I. (2004). Rab7 is required for the normal progression of the autophagic pathway in mammalian cells. J Cell Sci , 117, 2687-2697.

[20] Fader, C. M, Sanchez, D. G, Mestre, M. B, & Colombo, M. I. and VAMP3/cellubrevin: two v-SNARE proteins involved in specific steps of the autophagy/multivesicular body pathways. Biochim Biophys Acta , 1793, 1901-1916.

[21] Moreau, K, Ravikumar, B, Renna, M, Puri, C, & Rubinsztein, D. C. (2011). Autophagosome precursor maturation requires homotypic fusion. Cell , 146, 303-317.

[22] Furuta, N, Fujita, N, Noda, T, Yoshimori, T, & Amano, A. (2010). Combinational soluble N-ethylmaleimide-sensitive factor attachment protein receptor proteins VAMP8 and Vti1b mediate fusion of antimicrobial and canonical autophagosomes with lysosomes. Mol Biol Cell , 21, 1001-1010.

[23] Aguilera, M. O, Beron, W, & Colombo, M. I. (2012). The actin cytoskeleton participates in the early events of autophagosome formation upon starvation induced autophagy. Autophagy 8:

[24] Klionsky DJ, Abdalla FC, Abeliovich H, Abraham RT, cevedo-Arozena A, Adeli K, Agholme L, Agnello M, Agostinis P, guirre-Ghiso JA, Ahn HJ, it-Mohamed O, it-Si-Ali S, Akematsu T, Akira S, Al-Younes HM, Al-Zeer MA, Albert ML, Albin RL, egre-Abarrategui J, Aleo MF, Alirezaei M, Almasan A, monte-Becerril M, Amano A, Amaravadi R, Amarnath S, Amer AO, ndrieu-Abadie N, Anantharam V, Ann DK, noopkumar-Dukie S, Aoki H, Apostolova N, Arancia G, Aris JP, Asanuma K, Asare NY, Ashida H, Askanas V, Askew DS, Auberger P, Baba M, Backues SK, Baehrecke

EH, Bahr BA, Bai XY, Bailly Y, Baiocchi R, Baldini G, Balduini W, Ballabio A, Bamber BA, Bampton ET, Banhegyi G, Bartholomew CR, Bassham DC, Bast RC, Jr., Batoko H, Bay BH, Beau I, Bechet DM, Begley TJ, Behl C, Behrends C, Bekri S, Bellaire B, Bendall LJ, Benetti L, Berliocchi L, Bernardi H, Bernassola F, Besteiro S, Bhatia-Kissova I, Bi X, Biard-Piechaczyk M, Blum JS, Boise LH, Bonaldo P, Boone DL, Bornhauser BC, Bortoluci KR, Bossis I, Bost F, Bourquin JP, Boya P, Boyer-Guittaut M, Bozhkov PV, Brady NR, Brancolini C, Brech A, Brenman JE, Brennand A, Bresnick EH, Brest P, Bridges D, Bristol ML, Brookes PS, Brown EJ, Brumell JH, Brunetti-Pierri N, Brunk UT, Bulman DE, Bultman SJ, Bultynck G, Burbulla LF, Bursch W, Butchar JP, Buzgariu W, Bydlowski SP, Cadwell K, Cahova M, Cai D, Cai J, Cai Q, Calabretta B, Calvo-Garrido J, Camougrand N, Campanella M, Campos-Salinas J, Candi E, Cao L, Caplan AB, Carding SR, Cardoso SM, Carew JS, Carlin CR, Carmignac V, Carneiro LA, Carra S, Caruso RA, Casari G, Casas C, Castino R, Cebollero E, Cecconi F, Celli J, Chaachouay H, Chae HJ, Chai CY, Chan DC, Chan EY, Chang RC, Che CM, Chen CC, Chen GC, Chen GQ, Chen M, Chen Q, Chen SS, Chen W, Chen X, Chen X, Chen X, Chen YG, Chen Y, Chen Y, Chen YJ, Chen Z, Cheng A, Cheng CH, Cheng Y, Cheong H, Cheong JH, Cherry S, Chess-Williams R, Cheung ZH, Chevet E, Chiang HL, Chiarelli R, Chiba T, Chin LS, Chiou SH, Chisari FV, Cho CH, Cho DH, Choi AM, Choi D, Choi KS, Choi ME, Chouaib S, Choubey D, Choubey V, Chu CT, Chuang TH, Chueh SH, Chun T, Chwae YJ, Chye ML, Ciarcia R, Ciriolo MR, Clague MJ, Clark RS, Clarke PG, Clarke R, Codogno P, Coller HA, Colombo MI, Comincini S, Condello M, Condorelli F, Cookson MR, Coombs GH, Coppens I, Corbalan R, Cossart P, Costelli P, Costes S, Coto-Montes A, Couve E, Coxon FP, Cregg JM, Crespo JL, Cronje MJ, Cuervo AM, Cullen JJ, Czaja MJ, D'Amelio M, rfeuille-Michaud A, Davids LM, Davies FE, De FM, de Groot JF, de Haan CA, De ML, De MA, De T, V, Debnath J, Degterev A, Dehay B, Delbridge LM, Demarchi F, Deng YZ, Dengjel J, Dent P, Denton D, Deretic V, Desai SD, Devenish RJ, Di GM, Di PG, Di PC, az-Araya G, az-Laviada I, az-Meco MT, az-Nido J, Dikic I, nesh-Kumar SP, Ding WX, Distelhorst CW, Diwan A, Djavaheri-Mergny M, Dokudovskaya S, Dong Z, Dorsey FC (2012) Guidelines for the use and interpretation of assays for monitoring autophagy. Autophagy 8: 445-544.

[25] Morselli, E, Marino, G, Bennetzen, M. V, Eisenberg, T, Megalou, E, Schroeder, S, Cabrera, S, Benit, P, Rustin, P, Criollo, A, Kepp, O, Galluzzi, L, Shen, S, Malik, S. A, Maiuri, M. C, Horio, Y, Lopez-otin, C, Andersen, J. S, Tavernarakis, N, Madeo, F, & Kroemer, G. (2011). Spermidine and resveratrol induce autophagy by distinct pathways converging on the acetylproteome. J Cell Biol , 192, 615-629.

[26] Eisenberg, T, Knauer, H, Schauer, A, Buttner, S, Ruckenstuhl, C, Carmona-gutierrez, D, Ring, J, Schroeder, S, Magnes, C, Antonacci, L, Fussi, H, Deszcz, L, Hartl, R, Schraml, E, Criollo, A, Megalou, E, Weiskopf, D, Laun, P, Heeren, G, Breitenbach, M, Grubeck-loebenstein, B, Herker, E, Fahrenkrog, B, Frohlich, K. U, Sinner, F, Tavernarakis, N, Minois, N, Kroemer, G, & Madeo, F. (2009). Induction of autophagy by spermidine promotes longevity 1. Nat Cell Biol , 11, 1305-1314.

[27] Yuk, J. M, Yoshimori, T, & Jo, E. K. (2012). Autophagy and bacterial infectious diseas-
      es 2. Exp Mol Med , 44, 99-108.

[28] Chua, J, Vergne, I, Master, S, & Deretic, V. (2004). A tale of two lipids: Mycobacteri-
      um tuberculosis phagosome maturation arrest. Curr Opin Microbiol , 7, 71-77.

[29] Sturgill-koszycki, S, Schlesinger, P. H, Chakraborty, P, Haddix, P. L, Collins, H. L,
      Fok, A. K, Allen, R. D, Gluck, S. L, Heuser, J, & Russell, D. G. (1994). Lack of acidifi-
      cation in Mycobacterium phagosomes produced by exclusion of the vesicular proton-
      ATPase2. Science , 263, 678-681.

[30] Gutierrez, M. G, Master, S. S, Singh, S. B, Taylor, G. A, Colombo, M. I, & Deretic, V.
      (2004). Autophagy is a defense mechanism inhibiting BCG and Mycobacterium tu-
      berculosis survival in infected macrophages. Cell , 119, 753-766.

[31] Sakurai, A, Maruyama, F, Funao, J, Nozawa, T, Aikawa, C, Okahashi, N, Shintani, S,
      Hamada, S, Ooshima, T, & Nakagawa, I. (2010). Specific behavior of intracellular
      Streptococcus pyogenes that has undergone autophagic degradation is associated
      with bacterial streptolysin O and host small G proteins Rab5 and Rab7. J Biol Chem ,
      285, 22666-22675.

[32] Nakagawa, I, Amano, A, Mizushima, N, Yamamoto, A, Yamaguchi, H, Kamimoto, T,
      Nara, A, Funao, J, Nakata, M, Tsuda, K, Hamada, S, & Yoshimori, T. (2004). Autoph-
      agy defends cells against invading group A Streptococcus. Science , 306, 1037-1040.

[33] Yoshimori, T, & Amano, A. (2009). Group a Streptococcus: a loser in the battle with
      autophagy5. Curr Top Microbiol Immunol , 335, 217-226.

[34] Mauthe, M, Yu, W, Krut, O, Kronke, M, Gotz, F, Robenek, H, & Proikas-cezanne, T.
      (2012). WIPI-1 Positive Autophagosome-Like Vesicles Entrap Pathogenic Staphylo-
      coccus aureus for Lysosomal Degradation. Int J Cell Biol 2012: 179207.

[35] Andrade, R. M, Wessendarp, M, Gubbels, M. J, Striepen, B, & Subauste, C. S. (2006).
      CD40 induces macrophage anti-Toxoplasma gondii activity by triggering autophagy-
      dependent fusion of pathogen-containing vacuoles and lysosomes. J Clin Invest , 116,
      2366-2377.

[36] Subauste, C. S. (2009). Autophagy in immunity against Toxoplasma gondii. Curr Top
      Microbiol Immunol , 335, 251-265.

[37] Orvedahl, A, Alexander, D, Talloczy, Z, Sun, Q, Wei, Y, Zhang, W, Burns, D, Leib, D.
      A, & Levine, B. (2007). HSV-1 ICP34.5 confers neurovirulence by targeting the Beclin
      1 autophagy protein. Cell Host Microbe , 1, 23-35.

[38] Chaumorcel, M, Lussignol, M, Mouna, L, Cavignac, Y, Fahie, K, Cotte-laffitte, J, Ge-
      balle, A, Brune, W, Beau, I, Codogno, P, & Esclatine, A. (2012). The human cytomega-
      lovirus protein TRS1 inhibits autophagy via its interaction with Beclin 1. J Virol , 86,
      2571-2584.

[39] Sinha, S, Colbert, C. L, Becker, N, Wei, Y, & Levine, B. (2008). Molecular basis of the regulation of Beclin 1-dependent autophagy by the gamma-herpesvirus 68 Bcl-2 homolog M11. Autophagy , 4, 989-997.

[40] Dorn, B. R, & Dunn, W. A. Jr., Progulske-Fox A ((2001). Porphyromonas gingivalis traffics to autophagosomes in human coronary artery endothelial cells. Infect Immun , 69, 5698-5708.

[41] Belanger, M, Rodrigues, P. H, & Dunn, W. A. Jr., Progulske-Fox A ((2006). Autophagy: a highway for Porphyromonas gingivalis in endothelial cells. Autophagy , 2, 165-170.

[42] Swanson, M. S, & Fernandez-moreira, E. (2002). A microbial strategy to multiply in macrophages: the pregnant pause. Traffic , 3, 170-177.

[43] Amer, A. O, & Swanson, M. S. (2005). Autophagy is an immediate macrophage response to Legionella pneumophila. Cell Microbiol , 7, 765-778.

[44] Ghigo, E, Colombo, M. I, & Heinzen, R. A. (2012). The Coxiella burnetii Parasitophorous Vacuole.. Adv Exp Med Biol , 984, 141-169.

[45] Romano, P. S, Gutierrez, M. G, Beron, W, Rabinovitch, M, & Colombo, M. I. (2007). The autophagic pathway is actively modulated by phase II Coxiella burnetii to efficiently replicate in the host cell. Cell Microbiol , 9, 891-909.

[46] Beron, W, Gutierrez, M. G, Rabinovitch, M, & Colombo, M. I. (2002). Coxiella burnetii localizes in a Rab7-labeled compartment with autophagic characteristics. Infect Immun , 70, 5816-5821.

[47] Gutierrez, M. G, Vazquez, C. L, Munafo, D. B, Zoppino, F. C, Beron, W, Rabinovitch, M, & Colombo, M. I. (2005). Autophagy induction favours the generation and maturation of the Coxiella-replicative vacuoles. Cell Microbiol , 7, 981-993.

[48] Joshi, A. D, & Swanson, M. S. (2011). Secrets of a successful pathogen: legionella resistance to progression along the autophagic pathway. Front Microbiol 2: 138.

[49] Ogawa, M, Yoshimori, T, Suzuki, T, Sagara, H, Mizushima, N, & Sasakawa, C. (2005). Escape of intracellular Shigella from autophagy. Science , 307, 727-731.

[50] Kayath, C. A, & Hussey, S. El hN, Nagra K, Philpott D, Allaoui A ((2010). Escape of intracellular Shigella from autophagy requires binding to cholesterol through the type III effector, IcsB. Microbes Infect , 12, 956-966.

[51] Dortet, L, Mostowy, S, & Cossart, P. (2012). Listeria and autophagy escape: involvement of InlK, an internalin-like protein. Autophagy , 8, 132-134.

[52] Dortet, L, Mostowy, S, Samba-louaka, A, Gouin, E, Nahori, M. A, Wiemer, E. A, Dussurget, O, & Cossart, P. (2011). Recruitment of the major vault protein by InlK: a Listeria monocytogenes strategy to avoid autophagy. PLoS Pathog 7: e1002168.

[53] Jackson, W. T, & Giddings, T. H. Jr., Taylor MP, Mulinyawe S, Rabinovitch M, Kopito RR, Kirkegaard K ((2005). Subversion of cellular autophagosomal machinery by RNA viruses. PLoS Biol 3: e156.

[54] Berkova, Z, Crawford, S. E, Trugnan, G, Yoshimori, T, Morris, A. P, & Estes, M. K. (2006). Rotavirus NSP4 induces a novel vesicular compartment regulated by calcium and associated with viroplasms. J Virol , 80, 6061-6071.

[55] Checroun, C, Wehrly, T. D, Fischer, E. R, Hayes, S. F, & Celli, J. (2006). Autophagy-mediated reentry of Francisella tularensis into the endocytic compartment after cyto-plasmic replication. Proc Natl Acad Sci U S A , 103, 14578-14583.

[56] Wang, Y, Weiss, L. M, & Orlofsky, A. (2009). Host cell autophagy is induced by Toxo-plasma gondii and contributes to parasite growth. J Biol Chem , 284, 1694-1701.

[57] Andrews, N. W. (1994). From lysosomes into the cytosol: the intracellular pathway of Trypanosoma cruzi. Braz J Med Biol Res , 27, 471-475.

[58] Dvorak, J. A, & Howe, C. L. (1976). The attraction of Trypanosoma cruzi to vertebrate cells in vitro 4. J Protozool , 23, 534-537.

[59] Rodriguez, A, Samoff, E, Rioult, M. G, Chung, A, & Andrews, N. W. (1996). Host cell invasion by trypanosomes requires lysosomes and microtubule/kinesin-mediated transport. J Cell Biol , 134, 349-362.

[60] Tardieux, I, Webster, P, Ravesloot, J, Boron, W, Lunn, J. A, Heuser, J. E, & Andrews, N. W. (1992). Lysosome recruitment and fusion are early events required for trypa-nosome invasion of mammalian cells. Cell , 71, 1117-1130.

[61] Woolsey, A. M, Sunwoo, L, Petersen, C. A, Brachmann, S. M, Cantley, L. C, & Bur-leigh, B. A. kinase-dependent mechanisms of trypanosome invasion and vacuole ma-turation. J Cell Sci , 116, 3611-3622.

[62] Andrade, L. O, & Andrews, N. W. (2004). Lysosomal fusion is essential for the reten-tion of Trypanosoma cruzi inside host cells. J Exp Med , 200, 1135-1143.

[63] Andrews, N. W, & Whitlow, M. B. (1989). Secretion by Trypanosoma cruzi of a he-molysin active at low pH. Mol Biochem Parasitol , 33, 249-256.

[64] Ley, V, Robbins, E. S, Nussenzweig, V, & Andrews, N. W. (1990). The exit of Trypa-nosoma cruzi from the phagosome is inhibited by raising the pH of acidic compart-ments. J Exp Med , 171, 401-413.

[65] Andrews, N. W. (1994). From lysosomes into the cytosol: the intracellular pathway of Trypanosoma cruzi. Braz J Med Biol Res , 27, 471-475.

[66] Romano, P. S, Arboit, M. A, Vazquez, C. L, & Colombo, M. I. (2009). The autophagic pathway is a key component in the lysosomal dependent entry of Trypanosoma cru-zi into the host cell. Autophagy , 5, 6-18.

[67]   Romano, P. S, Arboit, M. A, Vazquez, C. L, & Colombo, M. I. (2009). The autophagic pathway is a key component in the lysosomal dependent entry of Trypanosoma cruzi into the host cell. Autophagy , 5, 6-18.

[68]   Romano, P. S, Cueto, J. A, Casassa, A. F, Vanrell, M. C, Gottlieb, R. A, & Colombo, M. I. (2012). Molecular and cellular mechanisms involved in the Trypanosoma cruzi/host cell interplay. IUBMB Life , 64, 387-396.

[69]   Alvarez, V. E, & Kosec, G. Sant'Anna C, Turk V, Cazzulo JJ, Turk B ((2008). Autophagy is involved in nutritional stress response and differentiation in Trypanosoma cruzi. J Biol Chem , 283, 3454-3464.

[70]   Cooper, A. M, Dalton, D. K, Stewart, T. A, Griffin, J. P, Russell, D. G, & Orme, I. M. (1993). Disseminated tuberculosis in interferon gamma gene-disrupted mice. J Exp Med , 178, 2243-2247.

[71]   Deretic, V, & Levine, B. (2009). Autophagy, immunity, and microbial adaptations. Cell Host Microbe , 5, 527-549.

[72]   Zhao, Z, Fux, B, Goodwin, M, Dunay, I. R, Strong, D, Miller, B. C, Cadwell, K, Delgado, M. A, Ponpuak, M, Green, K. G, Schmidt, R. E, Mizushima, N, Deretic, V, Sibley, L. D, & Virgin, H. W. (2008). Autophagosome-independent essential function for the autophagy protein Atg5 in cellular immunity to intracellular pathogens. Cell Host Microbe , 4, 458-469.

Chapter 4

Up-Regulation of Autophagy Defense Mechanisms in Mouse Mesenchymal Stromal Cells in Response to Ionizing Irradiation Followed by Bacterial Challenge

Nikolai V. Gorbunov, Thomas B. Elliott,
Dennis P. McDaniel, K. Lund, Pei-Jyun Liao,
Min Zhai and Juliann G. Kiang

Additional information is available at the end of the chapter

1. Introduction

Mesenchymal stroma along with epithelium, endothelium, reticuloendothelium, and lymphoid components is an essential constituent of tissue barriers that sustain immunochemical homeostatic interactions of tissue with internal and external environments. Thus, mesenchymal stroma protects the body from infections [1-8]. A breach of immune and structural integrity of tissue barriers under patho-physiological conditions such as complicated injury can lead to translocation of bacteria from different host-associated microbiomes and colonization of vital organs that can ultimately result in multiple organ failure and sepsis [9].

It is well documented that suppression of radiosensitive lymphoid and epithelial cells by ionizing irradiation results in impairment of tissue barriers and provokes bacterial translocation and sepsis leading to lethal outcome [10-12]. Under these circumstances one would expect increasing stress impact to the ubiquitously present and relatively radioresistant mesenchymal stromal components and their implication in host defense response [13]. This idea is supported by experimental and clinical observations indicating that injury can induce recruitment of mesenchymal stromal cells (MSCs) from bone marrow and promote their proliferative activity in order to re-constitute integrity of fractured tissue and mediate natural debridement [2, 5, 14, 15]. Moreover, it has been shown recently that transplanted MSCs can suppress experimental sepsis and can promote healing of radiation-induced cutaneous injury and survival from acute radiation syndrome [16-20]. All of this evidence suggests that MSCs play a crucial role in mitigation of systemic and local effects of tissue injury under different pathophysiological conditions. However, in the case of total body irradiation, the dynamics of MSC response to inflammatory stimuli can be skewed due to the cytotoxic effects of irradiation; but the mechanisms of remodeling of irradiated MSCs and their antimicrobial barrier capacity are not known and need to be delineated. *In vivo* assessment of stromal cell responses against bacteria is nearly impossible, because of (i) complexity of the architecture of the mesenchymal network in tissues and (ii) the fact of lethal complications in the hematopoietic radiation syndrome occurring at so low a level of microorganisms that the responses are difficult to detect [3,10].

From this perspective our attention was attracted by the macroautophagy-lysosomal (autolysosomal) mechanism described recently *in vitro* in cultured mesenchymal fibroblastic stromal cells [13]. The autophagy/autolysosomal mechanism mediates cell secretory functions and biodegradation mechanisms implicated in phagocytosis and cell remodeling activated in response to damage to cell constituents, endoplasmic reticular (ER) stress, and protein misfolding [21-23]. Thus, the autolysosomal pathway is responsible for decomposition of damaged proteins and organelles as well as phagocytized bacteria and viruses and is considered to be a part of the innate defense mechanism [23- 25].

The dynamics of macroautophagy (hereafter referred to as autophagy) in mammalian cells are well described in recent reviews [22, 26-28]. It has been proposed that autophagy is initiated by the formation of the phagophore, followed by a series of steps, including the elongation and expansion of the phagophore, closure and completion of a double-membrane autophagosome (which surrounds a portion of the cytoplasm), autophagosome maturation through docking and fusion with an endosome (the product of fusion is defined as an amphisome) and/or lysosome (the product of fusion is defined as an autolysosome), breakdown and degradation of the autophagosome inner membrane and cargo through acid hydrolases inside the autolysosome, and, finally, release of the resulting macromolecules through permeases [22]. These processes, along with the drastic membrane traffic, are mediated by factors known as autophagy-related proteins (i.e., ATG-proteins) and the lysosome-associated membrane proteins (LAMPs) that are conserved in evolution [29]. The autophagic pathway is complex. To date there are over 30 ATG genes identified in mammalian cells as regulators of various steps of autophagy such as cargo recognition, autophagosome formation, etc. [22, 30]. The core molecular machinery is comprised of (i) components of signaling cascades, such as the ULK1 and ULK2 complexes and class III PtdIns3K complexes, (ii) autophagy membrane processing components such as mammalian Atg9 (mAtg9) that contributes to the delivery of membrane to the autophagosome as it forms, and (iii) two conjugation systems: the microtubule-associated protein 1 (MAP1) light chain 3 (i.e., LC3) and the Atg12–Atg5–Atg16L complex. The two conjugation systems are proposed to function during elongation and expansion of the phagophore membrane [22, 27, 30]. A conservative estimate of the autophagy network counts over 400 proteins, which, besides the ATG-proteins, also including stress-response factors, cargo adaptors, and chaperones such as p62/SQSTM1 and heat shock protein 70 (HSP70) [23, 27, 30, 32, 33-35].

Autophagy is considered as a cytoprotective process leading to tissue remodeling, recovery, and rejuvenation. However, under circumstances leading to mis-regulation of the autolysosomal pathway, autophagy can eventually cause cell death, either as a precursor of apoptosis in apoptosis-sensitive cells or as a result of destructive self-digestion [36].

We hypothesized that: (i) MSCs enable activation of the autophagy pathway in response to ionizing irradiation; (ii) this mechanism is a part of adaptive remodeling essential for recovery of MSCs from the radiation-induced injury; and (iii) activation of autophagy in the irradiated MSCs can be potentiated by a challenge with Gram-negative or Gram-positive bacteria, e.g., *Escherichia coli* or *Staphylococcus epidermidis*, in order to sustain the MSC phagocytic antibacterial functions. The objective of the current chapter is to provide evidence to substantiate the proposed hypothesis.

2. Hypothesis test: Experimental procedures and technical approach

2.1. Mouse bone marrow Mesenchymal Stromal Cells (MSCs)

The cultures of MSCs were established and expanded as described previously [13]. Phenotype, proliferative activity, and colony-forming ability of the cells were monitored by flow cytometry and immunofluorescence imaging using positive markers for MSCs, i.e., CD44 and Sca1 [13].

2.2. Irradiation of MSCs and challenge with bacteria

2.2.1. MSC irradiation

MSC irradiation with gamma-photons was conducted using the ^{60}Co source in the Armed Forces Radiobiology Research Institute. The range of the applied doses was from 1 Gy through 12 Gy at a dose rate of 0.4 Gy/min. Dosimetry was performed using the alanine/electron paramagnetic resonance system. Calibration of the dose rate with alanine was traceable to the National Institute of Standards and Technology and the National Physics Laboratory of the United Kingdom. The irradiated cells were given a 24 h rest and then were subjected to either analyses or a challenge with *S. epidermidis* or *E. coli*.

2.2.2. Challenge of MSCs with bacteria

Irradiated and non-irradiated MSC cultures (~90% confluency) were challenged with either *S. epidermidis* or *E. coli* (5×10^7 bacteria/ml) for 1-3 h in antibiotic-free medium. For assessment of the cellular alteration during a period ≥ 3 h, the incubation medium was replaced with fresh medium containing penicillin and streptavidin antibiotics.

2.3. Cell analyses

Cell analysis for (i) the radiation-induced DNA double-strand breaks, viability, pro-apoptotic alterations, MSC proliferative activity, integrity of cell monolayers, and colony-forming activity; (ii) bacterial growth suppression, (iii) bacterial phagocytosis and autophagy (ATG), and (iv) response of stress-proteins, were conducted using flow cytometry techniques, fluorescence confocal imaging, protein immunoblotting, bright-field microscopy, and transmission electron microscopy (TEM).

The flow cytometry–based assessments of (i) the radiation-induced DNA double-strand breaks; (ii) proliferative activity; and (iii) cell viability were conducted using, respectively, the H2A.X phosphorylation assay kit (Cell Signaling Solutions, Temecula, CA); Click-iT® EdU Cell Proliferation Assay Kit, which utilizes a modified nucleoside, EdU (5-ethynyl-2′-deoxyuridine) that, in turn, is incorporated during *de novo* DNA synthesis in a quick-click chemistry reaction] (Life Technologies Corp., Grand Island, NY); and the CYTOX® Blue stain (Life Technologies Corp., Grand Island, NY).

The radiation-induced apoptotic response in MSCs was determined by immunoblot analysis of caspase-3, a marker of apoptosis.

The data presented in Fig. 1 indicate that the MSC cultures displayed a high integrity and survival from damage produced by irradiation with doses 1 Gy – 12 Gy. The irradiated cells challenged with bacteria were also able to sustain integrity of confluent monolayers (Fig. 1). The treated cells did not manifest signs of pro-apoptotic alterations. Moreover, MSCs challenged with *E. coli* and *S. epidermidis* at 24 h following irradiation (8 Gy) were able to suppress the bacterial growth by 1.4-fold and 1.8-fold, respectively (not shown).

2.4. Analysis of the cell proteins

Proteins from MSCs were extracted in accordance with the protocol described previously [12]. Aliquots of proteins were resolved on SDS-polyacrylamide slab gels (NuPAGE 4-12% Bis-Tris; Invitrogen, Carlsbad, CA). After electrophoresis, proteins were blotted onto a PDVF membrane and the blots were incubated with antibodies (1 µg/ml) raised against MAP LC3, Lamp1, p65(NFκB), HSP70, Sirt3a, SUMO1, and actin (Abcam, Santa Cruz Biotechnology Inc., LifeSpan Biosciences, Inc., eBiosciences) followed by incubation with species-specific IgG peroxidase conjugate.

2.5. Immunofluorescent staining and image analysis

MSCs (5 specimens per group) were fixed in 2% paraformaldehyde, processed for immunofluorescence analysis and analyzed with fluorescence confocal microscopy (30). Normal donkey serum and antibody were diluted in phosphate-buffered saline (PBS) containing 0.5% BSA and 0.15% glycine. Any nonspecific binding was blocked by incubating the samples with purified normal donkey serum (Santa Cruz Biotechnology, Inc., Santa Cruz, CA) diluted 1:20. Primary antibodies were raised against MAP LC3, Lamp1, p62/SQSTM1, p65(NFκB), FoxO3a, Tom 20. That was followed by incubation with secondary fluorochrome-conjugated antibody and/or streptavidin-AlexaFluor 610 conjugate (Molecular Probes, Inc., Eugene OR), and with Heochst 33342 (Molecular Probes, Inc., Eugene OR) diluted 1:3000. Secondary antibodies used were AlexaFluor 488 and AlexaFluor 594 conjugated donkey IgG (Molecular Probes Inc., Eugene OR). Negative controls for nonspecific binding included normal goat serum without primary antibody or with secondary antibody alone. Five confocal fluorescence and DIC images of crypts (per specimen) were captured with a Zeiss LSM 710 microscope. The immunofluorescence image analysis was conducted as described previously [12].

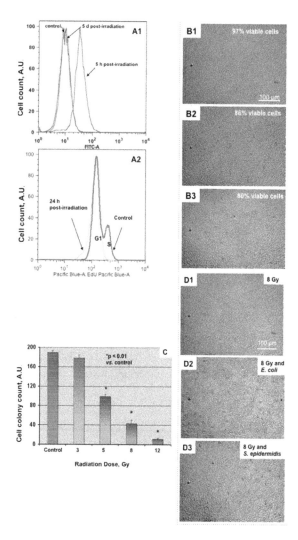

Figure 1. Functional ability of MSCs subjected to ionizing irradiation and bacterial challenge. A1. Analysis of radiation-induced DNA double-strand breaks with flow cytometry assay of the phosphorylated H2A.X (γ-H2A.X). Conditions: Control (in blue line) and irradiated MSCs were analyzed at 5 h (brown line) and 5 d (pink line) after 8-Gy irradiation. A2. Analysis of the radiation-induced suppression of MSC proliferative activity with flow cytometry assay of incorporated EdU, a modified nucleoside. The S-phase cell population was absent after irradiation Conditions: Control (red line) and irradiated (8 Gy, blue line) MSCs were analyzed 24 h after 8-Gy irradiation. A.U. is % of maximal cell count per channel. B. Bright-field microscopy analysis of effect of ionizing irradiation on ability of MSCs to form confluent

monolayers. Panel B1, control; Panel B2, 8-Gy irradiation; Panel B3, 12-Gy irradiation. Conditions: Images were captured 24 h after irradiation. Panel C. Analysis of the radiation-induced suppression of MSC colony-forming ability. Conditions: MSCs were harvested 24 h after irradiation and 200 MSCs from each radiation dose sample were aliquoted to Petri dishes and cultivated for 10 days. Panels D. Bright-field microscopy analysis of effects of bacterial challenge on ability of the irradiated MSCs shown in panel "B2" to form confluent monolayers. Panel D1 is after irradiation only (8 Gy), Panel D2 is same as "Panel D1" but after challenge with *E. coli*; Panel D3 is same as "Panel D1" but after challenge with *S. epidermidis*. Conditions: Images were captured 24 h after the bacterial challenge.

2.6. Transmission Electron Microscopy (TEM)

MSCs in culture were fixed in 4% formaldehyde and 4% glutaraldehyde in PBS overnight, post-fixed in 2% osmium tetroxide in PBS, dehydrated in a graduated series of ethanol solutions, and embedded in Spurr's epoxy resin. Blocks were processed as described previously [12,13]. The sections of embedded specimens were analyzed with a Philips CM100 electron microscope.

2.7. Statistical analysis

Statistical significance was determined using Student's *t*-test for independent samples. Significance was reported at a level of p<0.05.

3. Role of autophagy in adaptive response of MSCs to radiation injury and phagocytosis of *S. epidermidis* and *E. coli*

3.1. Alterations in the MSC stress-response-proteins following irradiation and bacterial challenge

The 8-Gy irradiation resulted in substantial DNA double-strand breaks in MSCs detectable with the γ-H2AX assay at 5 h post-exposure (Fig. 1A, brown line). This effect disappeared at 5 d post-irradiation recovery (Fig. 1A1, pink line). The observed DNA damage was accompanied by suppression of the cell proliferative activity determined with Click-iT® EdU Cell Proliferation Assay. As shown in Fig. 1A2 (red line), in control groups the cell populations were represented by the cells in both G1 and S phases. Following irradiation, the entire cell population was in G1 phase. The data presented in Fig. 1B indicate that the MSC cultures displayed a high integrity and survival from damage produced by irradiation at doses ranging from 1 Gy to 12 Gy, but that their ability to form colonies was reduced in a radiation dose-dependent manner (Fig. 1C).

The irradiated cells challenged with bacteria were also able to sustain integrity of confluent monolayers (Fig. 1D). These cells did not manifest signs of pro-apoptotic alterations. Moreover, MSCs challenged with *E. coli* and *S. epidermidis* 24 h after irradiation (8 Gy) were able to suppress the bacterial growth by 1.4-fold and 1.8-fold, respectively (not shown).

The data presented in Fig. 1 indicate that, despite radiation-produced damage and suppression of proliferative activity, MSCs demonstrated substantial radioresistance and absence of significant apoptotic and necrotic transformations in a wide range of radiation doses, i.e., 1-12

Gy. Interactive investigation of the stress-response factors implicated in cell survival may be important for the development of effective therapies for radiation injury (RI).

According to a current paradigm, the general stress responses involve conserved signaling modules that, in turn, are interconnected to the cellular adaptive mechanisms [37]. It is suggested that the stress due to molecular and organelle damage, impact of pro-oxidants, and infections triggers a cascade of responses attributed to specific sensitive transcriptional and post-transcriptional mechanisms mediating inflammation, antioxidant response, adaptation, and remodeling [36-40]. Ionizing radiation (IR) *per se* stimulates signaling cascades mediated by transcription factors and pathways that are believed to play a central role in protective response(s) to the molecular and subcellular damage and the oxidative stress. They include (but are not limited to) a battery of thiol-containing redox-response elements, redox-sensitive transcription factors such as nuclear factor-kappa B (NFκB) and forkhead box O3a (FoxO3a), stress-response adaptors such as the chaperone heat-shock protein 70 (HSP70) and NAD$^+$-dependent deacetylase sirtuin-3 (Sirt3), and activators of the autolysosomal degradation. Overall, these effector systems are crucial in maintaining homeostasis, which is altered due to damage to the cell constituents [33, 40-46]. It should be noted that, while the role of the IR-induced NFκB response in cell survival is well communicated, HSP70, the mitochondrial Sirt3, and FoxO3a are relatively newly-determined players implicated in adaptive mechanisms [43-48]. Thus, it has recently been observed that HSP70 and Sirt3 can sustain cell radioresistance and antioxidant capacity of mitochondria respectively; and that FoxO3a can promote cell survival by inducing the expression of antioxidant enzymes, autophagy, and factors involved in cell cycle withdrawal, such as the cyclin-dependent kinase inhibitor (CKI) p27 [33, 44-48].

Although the transcription factors NFκB and FoxO3a are normally sequestered in the cytoplasm, ionizing irradiation, bacterial products, pro-inflammatory effectors, and oxidative stress can stimulate their nuclear translocation and DNA-binding activity [13, 42, 43]. NFκB and FoxO3a are known to regulate numerous genes, including autophagy genes, and therefore, could link responses to IR and bacterial challenge with up-regulation of autolysosomal activity [13, 40, 42, 43, 45]. We do not, however, exclude implication of stress-induced adaptors and chaperones such as the heat-shock proteins (HSPs). HSP70, in particular, was shown to promote cell radioresistance and can regulate autophagy [33, 46]. Therefore, we assumed that a battery of stress-sensitive mechanisms mediated by survival factors such as NFκB, FoxO3a, Sirt3, and HSP70 are involved in an adaptive response of MSCs to IR and bacterial challenge.

Immunoblotting analysis of stress-response proteins presented in Fig. 2A indicates that control MSCs had relatively high amounts of constitutively expressed HSP70 and (p65)NFκB and a detectable amount of Sirt3. These basal levels did not significantly change at 24 h following 8-Gy irradiation. A slight increase in HSP70, (p65)NFκB, and Sirt3 occurred only after 12-Gy irradiation. Up-regulation of Casp-3 was not detected in the irradiated MSCs (Fig. 2A), which suggested the absence of pro-apoptotic alterations. Additional bacterial challenge of the 8-Gy irradiated MSCs did not compromise their viability (Fig. 1D) and did not affect the profile of the stress-proteins, except that IR induced a significant increase in Sirt3, a mitochondrial stress-response protein, and MMP3, the type 3 matrix metalloproteinase, essential for remodeling of extracellular matrix (Fig. 2B).

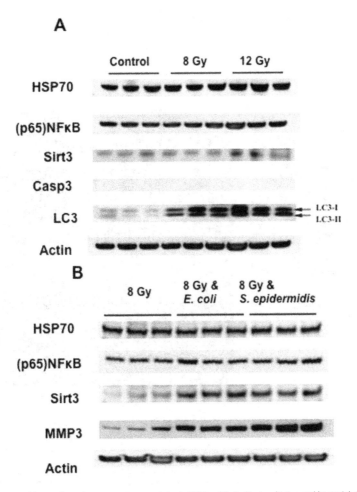

Figure 2. Immunoblot analysis of stress-response proteins in MSCs subjected to irradiation and bacterial challenge. A. MSCs, control and irradiated with 8 Gy and 12 Gy. Conditions: MSCs were harvested 24 h after irradiation then lysed and subjected to immunoblot analysis for stress-response proteins (HSP70, NFκB, and Sirt3), autolysosomal proteins (LC3-1 and LC3-II), and pro-apoptotic protein Caspase-3. B. MSCs irradiated with 8 Gy were challenged with either *E. coli* or *S. epidermidis*. Conditions: Irradiated MSCs were challenged with approximately 5x10⁷ bacteria /ml for 3 h in MesenCult Medium (without antibiotics). The cells were harvested and lysed 24 h after challenge. The protein lysates were subjected to immunoblot analysis for stress-response proteins (HSP70, NFκB, Sirt3, and MMP3).

Although in these experiments we did not observe a significant alteration of the amount of (p65)NFκB (Fig. 2B), the response of NFκB to IR was characterized by re-compartmentalization of (p65)NFκB resulting in an increase in its nuclear fraction (Fig. 3A). It should be noted that pre-incubation of the cells with pyrrolidine dithiocarbamate, an inhibitor of NFκB transloca-tion, or wortmannin, an inhibitor of autophagy, resulted in development of pro-apoptotic alterations and loss of confluency after irradiation (not shown). Immunofluorescence imaging of spacial localization of FoxO3a in MSCs (Fig. 3B) indicated that FoxO3a response to IR was associated with an increase in its nuclear fraction in a manner similar to (p65)NFκB.

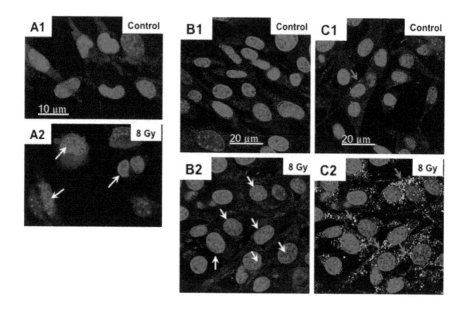

Figure 3. Confocal immunofluorescence imaging of nuclear translocation of NFκB and FoxO3a and activation of autophagy in MSCs subjected to 8-Gy irradiation. A. Projections of NFκB (red) in MSCs: A1, control; A2, 24 h after irradiation. Increase of nuclear fraction of p65 subunit of NFκB was observed in the irradiated cells due to transactivation of NFκB (indicated with white arrows). B. Projections of FoxO3a (red) in MSCs: B1, control; B2, 24 h after irradiation. Increase of nuclear fraction of FoxO3a was observed in the irradiated cells due to transactivation of FoxO3a (indicated with white arrows). C. Projections of LC3-positive autophagy vacuoles (green) in MSCs: C1, control; C2, 24 h after irradiation. A massive accumulation of autophagosomes occurred in irradiated MSCs (indicated with red arrows). Counterstaining of nuclei was with Hoechst 33342 (blue channel). The confocal images were taken with pinhole setup to obtain 0.5 μm Z-sections.

3.2. Autophagy—Autolysosomal response and secretory-activity in the irradiated MSCs subjected to bacterial challenge

The autophagy-autolysosomal pathway is considered to be an evolutionarily developed pro-survival mechanism, the purpose of which is to remove damaged and misfolded proteins, compromised organelles, and pathogens including bacteria [12, 21, 29, 34, 40]. A key step in autophagosome biogenesis is the conversion of light-chain protein 3 type I (LC3-I, also known as ubiquitin-like protein, Atg8) to type II (LC3-II). The conversion occurs via the cleavage of the LC3-I carboxyl terminus by a redox-sensitive Atg4 cysteine protease. The subsequent binding of the modified LC3-I to phosphatidylethanolamine, i.e., process of lipidation of LC3-1, on the isolation membrane, as it forms, is mediated by E-1- and E-2-like enzymes Atg7 and Atg3 [22, 26, 27, 40, 49]. Therefore, conversion of LC3-I to LC3-II and formation of LC3-positive vesicles are considered to be a marker of activation of autophagy [22, 26, 27, 33, 40]. Notably, a growing body of reports suggests implication of FoxO3a and HSP-70 in regulation of LC3 expression and translocation [32, 43, 45].

A line of evidence suggests that autophagy is a more selective process than the "bulk process" as it was originally defined [40, 49]. The discovery and characterization of autophagic adapters like p62/sequestrosome 1 (SQSTM1) and NBR1 (neighbor of BRCA1 gene 1), and target-ubiquitination with small ubiquitin-like modifier 1 (SUMO1) has provided mechanistic insight into this process. p62/SQSTM1 and NBR1 are both selectively degraded by autophagy and are able to act as cargo receptors for degradation of ubiquitinated/sumoylated substrates. A direct interaction between these autophagic adapters and the autophagosomal marker protein LC3-II, mediated by a so-called LIR (LC3-interacting region) motif, and their inherent ability to polymerize or aggregate, as well as their ability to specifically recognize substrates, are required for efficient selective autophagy [40, 49].

We hypothesized that autophagy and xenophagy (i.e., selective degradation of foreign pathogens by autophagy) can be implicated in the pro-survival response of MSCs to IR-related damage and bacterial challenge. To address this hypothesis we conducted immuno-blotting and immunofluorescence confocal imaging of autophagy MAP (LC3) protein, lysosomal LAMP1 and SUMO1 in MSCs after irradiation and challenge with *E. coli* and *S. epidermidis*.

The immunoblotting analysis of MSC proteins revealed a drastic increase in LC3-I and LC3-II (compared to control) at 24 h following 8-Gy and 12-Gy irradiation (Fig. 2A). These results suggested that, indeed, the autophagy MAP (LC3) pathway is implicated in that MSC response to IR-induced injury. In contrast to MAP (LC3), we did not observe a substantial increase in HSP-70. This was most likely due to relatively high background expression of this stress-response protein in the cells (Fig. 2A).

The above immunoblotting results were corroborated by the immunofluorescence confocal image analysis of the LC3 protein in MSCs. Thus, the data presented in Fig. 3C suggest that up-regulation of LC3-I/LC3-II proteins in the 8-Gy-irradiated cells was associated with massive formation of the LC3-positive vesicles which are well-documented to be features of autophagy [12, 13, 33]. The further TEM-assessment of the 8-Gy irradiated MSCs (in comparison with controls) revealed the presence of multiple vacuoles, which were formed by double-layer membrane and sequestered constituents of different densities (Figs. 4 A-C). Some of these vacuoles can be identified as secretory autolysosomes by the presence of multilamellar structures (most likely fibers of collagen) released extracellularly (Figs. 4 D and E), while others contained fractured organelles including mitochondria (Figs. 4 C and F).

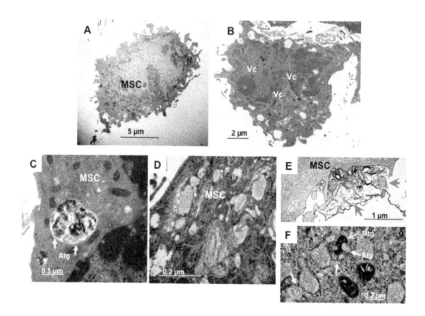

Figure 4. Transmission electron (TEM) analysis of autolysosomal vacuoles in irradiated MSCs. Panel A: A control MSC. Panel B: Irradiated MSC. A massive formation of different density autolysosomal vacuoles occurred after irradiation (indicated with pink arrows). Specimens were fixed 24 h after irradiation with 8 Gy gamma-photons. Panel C: Autolysosome sequestering cellular constituents (indicated with yellow arrows) in an irradiated MSC. Panel D: Formation of secretory autolysosomes containing multilamellar structures (indicated with red arrows) in an irradiated MSC. Panel E: Extracellular secretion of multilamellar structures from an irradiated cell. Panel F: Autolysosome sequestering mitochondria, e.g., mitophagy, (indicated with blue arrow) in an irradiated MSC. Abbreviations: "Vc", vacuoles; "Atg", autophagosomes/autolysosomes; "Mtg", mitophagy.

In our recent research we demonstrated that intact MSCs are able to up-regulate autophagy in response to challenge with *E. coli* and employ this mechanism for inactivation of the microorganisms [13]. The data presented in this report showed that the irradiated MSCs retained their ability to phagocytise bacteria in a manner similar to that of non-irradiated MSCs (Fig. 5). Indeed, sequestration and degradation of *E. coli* and *S. epidermidis* in the MSC vacuoles, constituted by characteristic autophagosomal membranes, was observed at 5 h after bacterial challenge of both non-irradiated and irradiated MSCs (Figs. 5 A, B, D, and E).

The immunofluorescence confocal image analysis of the irradiated MSCs challenged with bacteria showed that the vacuoles containing bacteria were LC3-positive and that this LC3 immunoreactivity was co-localized with immunoreactivity to LAMP1, a marker of lysosomes, indicating presence of fusion of autophagosomes with lysosomes, i.e., formation of autolysosomes (Fig. 6). This increase in autolysosomal activity was accompanied by accumulation of the proteins LC3-II, a marker of up-regulation of autophagy, LAMP1, and p62/SQSTM1, a target adaptor (Figs. 7A and C). Meanwhile, the level of SUMO1, a target modifier protein, in the cells decreased after bacterial challenge (Fig. 7A). Interestingly, irradiation and bacterial challenge resulted in up-regulation of the factors responsible for modification of extracellular matrix, such as collagen III, MMP3, and MMP13, i.e., the collagenase-3, (Fig. 7C), indicating that the stress-response aimed at multiple targets including extracellular ones.

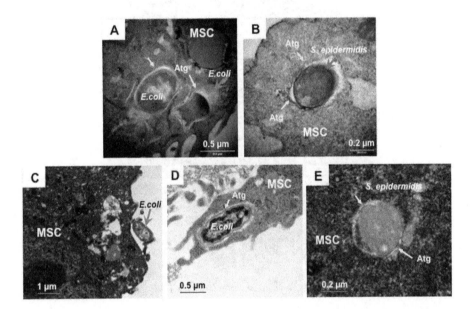

Figure 5. Assessment of phagocytosis and autophagy/autolysosomal processing of *E. coli* or *S. epidermidis* in irradiated MSCs with transmission electron microscopy. Panels A and B: Autolysosomal degradation of phagocytized *E. coli* and *S. epidermidis* in control MSCs. Autophagosome (ATG) membranes are indicated with yellow arrows. Conditions: Control MSCs were challenged with ~5x10[7] bacteria /ml for 3 h as indicated in *Methods*. The cells were harvested and fixed for TEM 5 h after challenge. Panels C, D, and E: TEM micrographs obtained from the 8-Gy irradiated cells. C - Engulfing and up-take of *E. coli* (pink arrows) by the cell plasma membrane extrusions (black arrows). D - Autolysosomal degradation of phagocytized *E. coli*. E - Autolysosomal degradation of phagocytized *S. epidermidis*. Atg, autophagosomes/autolysosomes. Conditions: 8-Gy irradiated MSCs were challenged with ~5x10[7] bacteria/ml for 3 h as indicated in *Methods*. The cells were harvested and fixed for TEM 5 h after challenge.

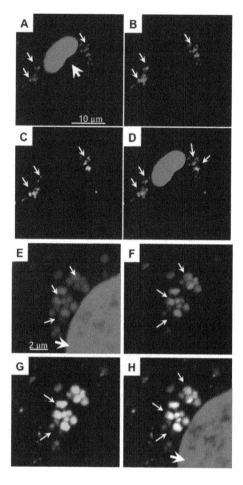

Figure 6. Confocal immunofluorescence imaging of autolysosomal sequestration of *E. coli* and *S. epidermidis* microorganisms phagocytized by MSCs irradiated at 8-Gy. Panel A – (Blue channel). *E. coli* (small arrows) and MSC nuclear DNA (large arrow) are indicated. Panel B – (Red channel). Spatial localization of LAMP1 is indicated with arrows. Panel C – (Green channel). Spatial localization of LC3 is indicated with arrows. Panel D – Overlay of images appeared in the blue, red, and green channels and presented in panels A, B, and C, respectively. Spatial co-localization of LAMP1, LC3, and *E. coli* DNA is indicated with arrows. Conditions: 8-Gy irradiated MSCs were challenged with ~5x10⁷ *E. coli*/ml for 3 h. The cells were fixed 24 h after challenge. The fixed cells were subjected to immunofluorescence analysis for autolysosomal proteins. Panel E – (Blue channel). *S. epidermidis* (small arrows) and MSC nuclear DNA (large arrow) are indicated. Panel F – (Red channel). Spatial localization of LAMP1 is indicated with arrows. Panel G – (Green channel). Spatial localization of LC3 is indicated with arrows. Panel H – Overlay of images appeared in the blue, red, and green channels and presented in panels E, F, and G, respectively. Spatial co-localization of LAMP1, LC3, and *S. epidermidis* DNA is indicated with arrows. Conditions: 8-Gy irradiated MSCs were challenged with ~5x10⁷ *S. epidermidis*/ml for 3 h. The cells were fixed 24 h after challenge. The fixed cells were subjected to immunofluorescence analysis for autolysosomal proteins. Counterstaining of nuclei was with Hoechst 33342 (blue channel). The confocal images were taken with pinhole setup to obtain 0.5 μm Z-sections.

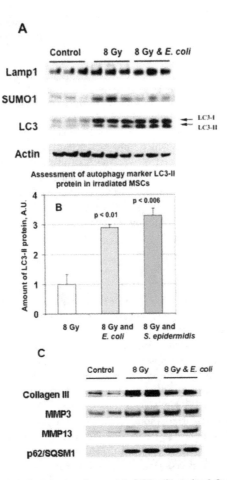

Figure 7. Immunoblot assessment of autolysosomal response in MSCs subjected to 8-Gy irradiation followed by challenge with either *E. coli* or *S. epidermidis*. Panel A. Representative immunoblotting bands of SUMO1, LC3, and LAMP1. Note that irradiated MSCs were challenged with *E. coli*. Panel B. Densitometry histograms of LC3-II bands of the immunoblots of proteins from the irradiated MSCs subjected to challenge with *E. coli* or *S. epidermidis*. The presented bars indicate the relative density of LC3-II protein (normalized to density of actin bands). The statistical significance was determined by Student's *t*-test (n=3). Panel C. Representative immunoblots of collagen III, MMP3, MMP13, and p62/SQSM1. Note that irradiated MSCs were challenged with *E. coli*. Conditions: Irradiated MSCs were challenged with approximately 5x10^7 bacteria/ml for 3 h as indicated in *Methods*. The cells were harvested and lysed 24 h after challenge.

Various cells eliminate bacterial microorganisms by autophagy, and this elimination is in many cases crucial for host resistance to bacterial translocation. Targeting of microorganisms can occur outside of the host cells in extracellular matrix by different defense mechanisms, such as the cell-produced oxidative burst, nitric oxide, antibacterial peptides, and extracellular traps [50]. The data presented in the present report (Figs. 2, 4, 5, 7) suggest that MSCs can employ the autophagy mechanism to modify extracellular matrix by releasing collagen and matrix metalloproteases in order to increase efficacy of extracellular entrapment, uptake, and further phagocytosis of the microorganisms.

Recent observations suggest that autophagosomes do not form randomly in the cytoplasm, but rather sequester the bacteria selectively [23, 49, 51]. Therefore, autophagosomes that engulf microbes are sometimes much larger than those formed during degradation of cellular organelles, suggesting that the elongation step of the autophagosome membrane is involved in bacteria-surrounding autophagy [13, 32]. This effect could be observed by comparison of profiles of the autophagosomes, which appeared in the irradiated MSCs before and after challenge with bacteria (Figs. 3 and 6). The mechanism underlying selective induction of autophagy at the site of microbe phagocytosis remains unknown. However, it is likely mediated by pattern recognition receptors, stress-response elements, adaptor proteins, e.g., p62/SQSTM1, and ubiquitin-like modifiers, which can target bacteria and ultimately recruit factors essential for the formation of autophagosomes [21, 22, 52].

4. Conclusion

Survival of multicellular organisms in a non-sterile environment requires a network of host defense mechanisms. The initial contact of pathogenic and opportunistic microorganisms with a host usually takes place at internal or external body surfaces. Microbial growth and translocation are controlled by multi-layer integrative tissue barriers that mediate innate defense mechanisms. Tissue injury compromises barrier function, and increases risk of infection and sepsis. Recent observations from our laboratory indicate that ubiquitous MSCs can modulate systemic responses to bacterial infection and support tissue repair and healing when recruited at sites of injury [2, 5, 9, 14-18]. However, these "compensatory" responses of MSCs can be skewed and suppressed after irradiation. To elucidate the role of autophagy in response of stromal cells to radiation injury and bacterial infection we irradiated cultured MSCs and challenged them with *S. epidermidis* or *E. coli*. Using this cell model we showed that (i) irradiation induced translocation of cytosol NF-kB and FoxO3a to the nucleus; (ii) irradiation and bacterial challenge induced increases in Sirt3 stress-response factors, LC3, MMP3, MMP13, collagen III, SUMO1, and p62/SQSM1 proteins; and (iii) the antibacterial defense response of the irradiated MSCs was characterized by extensive phagocytosis and inactivation of both *S. epidermidis* or *E. coli* in autolysosomes.

Our communication is the first report demonstrating a potential role of MSCs in sustaining antibacterial barrier functions of irradiated tissues. We postulate that effector mechanisms expressed by MSCs can contribute to the innate defense response to IR injury alone or, especially, when IR is combined with trauma.

Grants

This work was supported by AFRRI Intramural RAB2CF (to JGK) and NIAID YI-AI-5045-04 (to JGK). There are no ethical and financial conflicts in the presented work.

Disclaimer

The opinions or assertions contained herein are the authors' private views and are not to be construed as official or reflecting the views of the Uniformed Services University of the Health Sciences, Armed Forces Radiobiology Research Institute, the U.S. Department of Defense or the National Institutes of Health.

Acknowledgements

The authors thank Prof. J.S . Greenberger (UPMC/University of Pittsburgh Schools of the Health Sciences) for his criticism and encouragement in conducting of our investigation, and Ms. Dilber Nurmemet for her technical support.

Author details

Nikolai V. Gorbunov[1*], Thomas B. Elliott[2], Dennis P. McDaniel[3], K. Lund[3], Pei-Jyun Liao[1], Min Zhai[1] and Juliann G. Kiang[2,3*]

*Address all correspondence to: nikolaiv.gorbunov@gmail.com

1 The Henry M. Jackson Foundation, USA

2 Armed Forces Radiobiology Research Institute, USA

3 School of Medicine, Uniformed Services University of the Health Sciences, Bethesda, Maryland, USA

References

[1] Phan, S. H. Biology of fibroblasts and myofibroblasts. Proc Am Thorac Soc (2008). , 5, 334-7.

[2] Krebsbach, P. H, Kuznetsov, S. A, Bianco, P, & Robey, P. G. Bone marrow stromal cells: characterization and clinical application. Crit Rev Oral Biol Med (1999). , 10, 65-81.

[3] Winkler, I. G, Barbier, V, Wadley, R, Zannettino, A. C, Williams, S, & Lévesque, J. P. Positioning of bone marrow hematopoietic and stromal cells relative to blood flow *in vivo*: serially reconstituting hematopoietic stem cells reside in distinct nonperfused niches. Blood (2010). , 116, 375-85.

[4]   Zusman, I, Gurevich, P, & Ben-hur, H. Two secretory immune systems (mucosal and barrier) in human intrauterine development, normal and pathological. Int J Mol Med (2005). , 16, 127-33.

[5]   Powell, D. W, Pinchuk, I. V, & Saada, J. I. Chen Xin, and Mifflin RC. Mesenchymal Cells of the Intestinal Lamina Propria. Annu Rev Physiol (2011). , 73, 13-237.

[6]   Turner, H. L, & Turner, J. R. Good fences make good neighbors: Gastrointestinal mucosal structure. Gut Microbes (2010). , 1, 22-29.

[7]   Hooper, L. V, & Macpherson, A. J. Immune adaptations that maintain homeostasis with the intestinal microbiota. Nat Rev Immunol (2010). , 10, 159-69.

[8]   Louis, N. A, & Lin, P. W. The Intestinal Immune Barrier. Neoreviews (2009). ee190., 180.

[9]   Kirkup BC Jr, Craft DW, Palys T, Black C, Heitkamp R, Li C, Lu Y, Matlock N, McQueary C, Michels A, Peck G, Si Y, Summers AM, Thompson M, Zurawski DV.Traumatic wound microbiome workshop. Microb Ecol (2012). , 64, 37-50.

[10] Ledney, G. D, & Elliott, T. B. Combined injury: factors with potential to impact radiation dose assessments. Health Physics (2010). , 98, 145-152.

[11] Kiang, J. G, Jiao, W, Cary, L. H, Mog, S. R, Elliott, T. B, Pellmar, T. C, & Ledney, G. D. Wound trauma increases radiation-induced mortality by activation of iNOS pathway and elevation of cytokine concentrations and bacterial infection. Radiat Res (2010). , 173, 319-32.

[12] Gorbunov, N. V, Garrison, B. R, & Kiang, J. G. Response of crypt Paneth cells in the small intestine following total-body gamma-irradiation. Int J Immunopathol Pharmacol (2010). , 23, 1111-23.

[13] Gorbunov, N. V, Garrison, B. R, Zhai, M, Mcdaniel, D. P, Ledney, G. D, Elliott, T. B, & Kiang, J. G. Autophagy-Mediated Defense Response of Mouse Mesenchymal Stromal Cells (MSCs) to Challenge with *Escherichia coli*. In: Cai J and Wang RE (eds) Protein Interactions. Rijeka: InTech; (2012). , 23-44.

[14] Jackson, W. M, Alexander, P. G, Bulken-hoover, J. D, Vogler, J. A, Ji, Y, Mckay, P, Nesti, L. J, & Tuan, R. S. Mesenchymal progenitor cells derived from traumatized muscle enhance neurite growth. J Tissue Eng Regen Med (2012). doi:10.1002/term. 539.Epub ahead of print].

[15] Jackson, W. M, Aragon, A. B, Djouad, F, Song, Y, Koehler, S. M, Nesti, L. J, & Tuan, R. S. Mesenchymal progenitor cells derived from traumatized human muscle. J Tissue Eng Regen Med (2009). , 3, 129-38.

[16] Le Blanc K, Ringdén O.Immunomodulation by mesenchymal stem cells and clinical experience. J Intern Med (2007). , 262, 509-25.

[17] Lee, J. W, Fang, X, Gupta, N, Serikov, V, & Matthway, M. A. Allogeneic human mesenchymal stem cells for treatment of E. coli endotoxin-induced acute lung injury in the ex vivo perfused human lung. Proc Natl Acad Sci USA (2009). , 106, 16357-62.

[18] Nemeth, K, Mayer, B, & Mezey, E. Modulation of bone marrow stromal cell functions in infectious diseases by toll-like receptor ligands. J Mol Med (2010). , 88, 5-10.

[19] Akita, S, Akino, K, Hirano, A, Ohtsuru, A, & Yamashita, S. Mesenchymal stem cell therapy for cutaneous radiation syndrome. Health Phys (2010). , 98, 858-62.

[20] Lange, C, Brunswig-spickenheier, B, Cappallo-obermann, H, Eggert, K, Gehling, U. M, Rudolph, C, Schlegelberger, B, Cornils, K, Zustin, J, Spiess, A. N, & Zander, A. R. Radiation rescue: mesenchymal stromal cells protect from lethal irradiation. PLoS One (2011). e14486.

[21] Levine, B, Mizushima, N, & Virgin, H. W. Autophagy in immunity and inflammation. Nature (2011). , 469, 323-35.

[22] Yang, Z, & Klionsky, D. J. Eaten alive: a history of macroautophagy. Nat Cell Biol (2010). , 12, 814-22.

[23] Yano, T, & Kurata, S. Intracellular recognition of pathogens and autophagy as an innate immune host defence. J Biochem (2011). , 150, 143-9.

[24] Mizushima, N, Levine, B, Cuervo, A. M, & Klionsky, D. J. Autophagy fights disease through cellular self-digestion. Nature (2008). , 451, 1069-75.

[25] Klionsky, D. J. The Autophagy Connection. Dev Cell. (2010). , 19, 11-2.

[26] Tooze, S. A, & Yoshimori, T. The origin of the autophagosomal membrane. Nat Cell Biol (2010). , 12, 831-5.

[27] Weidberg, H, Shvets, E, & Elazar, Z. Biogenesis and cargo selectivity of autophagosomes. Annu Rev Biochem (2011). , 80, 125-56.

[28] Eskelinen, E. L. New insights into the mechanisms of macroautophagy in mammalian cells. Int Rev Cell Mol Biol (2008). , 266, 207-47.

[29] Eskelinen, E. L, & Saftig, P. Autophagy: a lysosomal degradation pathway with a central role in health and disease. Biochim Biophys Acta (2009). , 1793, 664-73.

[30] Mizushima, N, & Levine, B. Autophagy in mammalian development and differentiation. Nat Cell Biol (2010). , 12, 823-30.

[31] Kabeya, Y, Mizushima, N, Yamamoto, A, Oshitani-okamoto, S, Ohsumi, Y, & Yoshimori, T. LC3, GABARAP and GATE16 localize to autophagosomal membrane depending on form-II formation. J Cell Sci (2004). , 117, 2805-12.

[32] Behrends, C, Sowa, M. E, Gygi, S. P, & Harper, J. W. Network organization of the human autophagy system. Nature (2010). , 466, 68-76.

[33] Viiri, J, Hyttinen, J. M, Ryhänen, T, Rilla, K, Paimela, T, Kuusisto, E, Siitonen, A, Urt-ti, A, Salminen, A, & Kaarniranta, K. p/62 sequestosome 1 as a regulator of protea-some inhibitor-induced autophagy in human retinal pigment epithelial cells. Mol Vis (2010). , 16, 1399-414.

[34] Ryhänen, T, Hyttinen, J. M, Kopitz, J, Rilla, K, Kuusisto, E, Mannermaa, E, Viiri, J, Holmberg, C. I, Immonen, I, Meri, S, Parkkinen, J, Eskelinen, E. L, Uusitalo, H, Salmi-nen, A, & Kaarniranta, K. Crosstalk between Hsp70 molecular chaperone, lysosomes and proteasomes in autophagy-mediated proteolysis in human retinal pigment epi-thelial cells. J Cell Mol Med (2009). , 13, 3616-31.

[35] Behl, C. BAG3 and friends: co-chaperones in selective autophagy during aging and disease. Autophagy (2011). , 7, 795-8.

[36] Sridhar, S, Botbol, Y, Macian, F, & Cuervo, A. M. Autophagy and Disease: always two sides to a problem. J Pathol (2012). , 226, 255-73.

[37] Kültz, D. Molecular and evolutionary basis of the cellular stress response. Annu Rev Physiol (2005). , 67, 225-57.

[38] Burhans, W. C, & Heintz, N. H. The cell cycle is a redox cycle: linking phase-specific targets to cell fate. Free Radic Biol Med (2009). , 47, 1282-93.

[39] Baltimore, D. NF-κB is 25. Nat Immunol (2011). , 12, 683-5.

[40] Murrow, L, & Debnath, J. Autophagy as a Stress-Response and Quality-Control Mechanism: Implications for Cell Injury and Human Disease. Annu Rev Pathol. (2012). Oct 15. [Epub ahead of print]

[41] Wei, S. J, Botero, A, Hirota, K, Bradbury, C. M, Markovina, S, Laszlo, A, Spitz, D. R, Goswami, P. C, Yodoi, J, & Gius, D. Thioredoxin nuclear translocation and interac-tion with redox factor-1 activates the activator protein-1 transcription factor in re-sponse to ionizing radiation. Cancer Res (2000). , 60, 6688-95.

[42] Li, N, & Karin, M. Ionizing radiation and short wavelength UV activate NF-κB through two distinct mechanisms. Proc Natl Acad Sci USA (1998). , 95, 13012-7.

[43] Tsai, W. B, Chung, Y. M, Takahashi, Y, Xu, Z, & Hu, M. C. Functional interaction be-tween FOXO3a and ATM regulates DNA damage response. Nat Cell Biol (2008). , 10, 460-7.

[44] Aquila, D, Rose, P, Panno, G, Passarino, M. L, Bellizzi, G, & , D. Sirt3 gene expres-sion: a link between inherited mitochondrial DNA variants and oxidative stress. Gene (2012). , 497, 323-9.

[45] Rodriguez-rocha, H, Garcia-garcia, A, Panayiotidis, M. I, & Franco, R. DNA damage and autophagy. Mutat Res (2011). , 711, 158-66.

[46] Graner, M. W, Raynes, D. A, Bigner, D. D, & Guerriero, V. Heat shock protein 70-binding protein 1 is highly expressed in high-grade gliomas, interacts with multiple

heat shock protein 70 family members, and specifically binds brain tumor cell surfaces. Cancer Sci (2009). , 100, 1870-9.

[47] Miyamoto, K, Araki, K. Y, Naka, K, Arai, F, Takubo, K, Yamazaki, S, Matsuoka, S, Miyamoto, T, Ito, K, Ohmura, M, Chen, C, Hosokawa, K, Nakauchi, H, Nakayama, K, Nakayama, K. I, Harada, M, Motoyama, N, Suda, T, & Hirao, A. Foxo3a is essential for maintenance of the hematopoietic stem cell pool. Cell Stem Cell (2007). , 1, 101-12.

[48] Burhans, W. C, & Heintz, N. H. The cell cycle is a redox cycle: linking phase-specific targets to cell fate. Free Radic Biol Med (2009). , 47, 1282-93.

[49] Reggiori, F, Komatsu, M, Finley, K, & Simonsen, A. Selective types of autophagy. Int J Cell Biol (2012). Epub 2012 May 15.

[50] von Köckritz-Blickwede, M, & Nizet, V. Innate immunity turned inside-out: antimicrobial defense by phagocyte extracellular traps. J Mol Med (Berl) (2009). , 87, 775-83.

[51] Nakagawa, I, Amano, A, Mizushima, N, Yamamoto, A, Yamaguxhi, H, Kamimoto, T, Nara, A, Funao, J, Nakata, M, Tsuda, K, Hamada, S, & Yoshimori, T. Autophagy defends cells against invading Group A *Streptococcus*. Science (2004). , 306, 1037-40.

[52] Ichimura, Y, & Komatsu, M. Selective degradation of by autophagy. Semin Immunopathol (2010). , 62.

Autophagy in Neurodegenerative Diseases

Chapter 5

Neuronal Autophagy and Prion Proteins

Audrey Ragagnin, Aurélie Guillemain,
Nancy J. Grant and Yannick J. R. Bailly

Additional information is available at the end of the chapter

1. Introduction

Protein and organelle turnover is essential to maintain cellular homeostasis and survival. Removing and recycling cell constituents is achieved by autophagy in all cells, including neurons. Autophagy contributes to various physiological processes, such as intracellular cleansing, cellular homeostasis, development, differentiation, longevity, tumor suppression, elimination of invading pathogens, antigen transport to the innate and adaptive immune systems, and counteracting endoplasmic reticulum (ER) stress and diseases characterized by the accumulation of protein aggregates [1]. Autophagy plays a role in a number of infectious and inflammatory diseases, in addition to protein unfolding and misfolding diseases that lead to neuron, muscle and liver degeneration or heart failure [2-4]. Whereas autophagy has long been defined as a form of non-apoptotic, programmed cell death [5], recent findings suggest that autophagy functions primarily to sustain cells, and only defects in autophagy lead to cell death [6].

2. Autophagy in neuronal physiology

Autophagy was initially identified and characterized in a few cell types including neurons. The distinct vacuoles which feature this self-eating process were originally described at the ultrastructural level [7, 8]. The formation of autophagosomes was associated with chromatolysis of a restricted neuroplasm area, free of organelles, but filled with various types of vesicles [9]. The function of autophagy in mature neurons , however, is still debated. In comparison with other organs, rodent brains show high expression levels of autophagy-related (Atg) proteins and low levels of autophagy markers such as autophagosome number and LC3-II. Indeed, even under prolonged fasting conditions, the number of autophagosomes does not increase in neurons, probably because their nutrient supply from peripheral organs is maintained [10]. However, mice with CNS defects in their autophagic machinery exhibit neurological deficits, such as abnormal limb-clasping reflexes, locomotor ataxia, and lack of motor coordination, in addition to a significant loss of large pyramidal neurons in the cerebral cortex and Purkinje cells (PCs) in the cerebellar cortex [11-13].

Macroautophagy (hereafter referred to as autophagy) is initiated when a portion of the cytoplasm is sequestrated within a double-membrane organelle, the so-called autophagosome [14]. The autophagic machinery has been extensively detailed at the molecular level in a number of reviews including several chapters of this book [14-17]. Atg and several non-Atg proteins have been identified as regulators of key steps leading to the degradation of cytosolic components in lysosomes: initiation and nucleation of phagophores, expansion of autophagosomes, maturation of autophagosomes into amphisomes/autolysosomes, and execution of autophagic degradation [18]. The endosomal sorting complex required for transport (ESCRT) pathway functions in the sorting of transmembrane proteins into the inner vesicles of multivesicular bodies (MVB) during endocytosis. Also it is conceivably an essential part of the basal autophagy process in neurons because ubiquitin- or p62/SQSTM1 (p62)-labelled inclusions and autophagosomes accumulate in neurons deficient in ESCRT components [19]. Increasing evidence suggests that autophagy is regulated in a cell type-specific manner and as such autophagy may serve a distinct function in neurons and may show difference in the molecular machinery underlying basal autophagy (Fig. 1).

2.1. Axonal autophagy

In neurons, autophagy occurs in axons, suggesting that it may be uniquely regulated in this compartment and specifically adapted to local axonal physiology [20]. In primary dorsal root ganglion neurons, autophagosomes initiate distally in nerve terminals and mature during their transport toward the cell soma [21, 22]. In non-neuronal cells, the autophagosomal membrane has multiple possible origins, including endocytosed plasma membrane (amphisome), ER, mitochondria, and trans-Golgi membranes [18, 21, 23- 32]. In contrast, the origin of autophagosomes in the axons is likely to be restricted to the sources of membrane available in the terminals such as smooth ER and plasma membrane [33, 34], excluding rough ER and Golgi dictyosomes.

As observed in ultrastructural studies of axotomized neurons [9, 35], analysis of Purkinje cell (PC) degeneration in lurcher mutant GluRδ2^{Lc} [36] demonstrates autophagosomes in their axonal compartment. In this study, an excitotoxic insult mediated by GluRδ2^{Lc} triggered a rapid and robust accumulation of autophagosomes in dystrophic axonal swellings providing evidence that autophagy is induced in dystrophic terminals and that autophagosome biogenesis occurs in axons [37]. The molecular scenario underlying the initiation of this axonal autophagy is unclear. Liang et al. [38] suggested that autophagy in lurcher PCs could be directly activated by an interaction between the postsynaptic GluRδ2^{Lc}, nPIST and beclin 1 an important regulator of autophagy. Nevertheless, how activation of this postsynaptic signaling pathway in dendrites initiates autophagosome formation in axon compartments is uncertain. PC death that is correlated with early signs of autophagy appears to be independent of depolarization

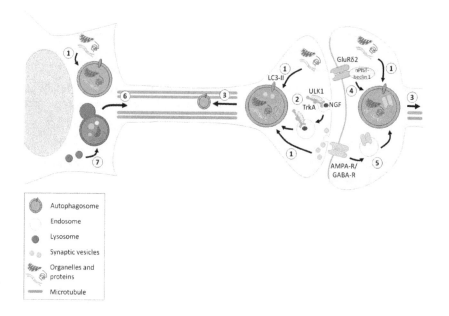

Figure 1. Physiological neuronal autophagy. Autophagy recycles synaptic components to sustain neuronal homeo-stasis and regulate synaptic plasticity and growth. **1.** Autophagic degradation of organelles, synaptic vesicles and pro-teins. **2.** ULK1-mediated autophagy of endocytosed NGF-bound TrkA receptors. **3.** Dynein-mediated retrograde transport of autophagosomes. **4.** GluRδ2 activation of beclin 1-dependent autophagy via nPIST. **5.** Targeting of post-synaptic receptors to autophagosomes via endocytosis. **6.** Kinesin-dependent anterograde transport of autophago-somes. **7.** Formation of autophagolysosomes by fusion of autophagosomes with lysosomes.

in the heteroallelic mutant Lurcher/hotfoot bearing only one copy of the lurcher allele and no wild-type GRID2 [39]. However, in the lurcher mutant bearing only one copy of the lurcher allele and one copy of the wild-type GRID2 allele, the leaky channel of GluRδ2^{Lc} depolarizes the neuron and this could transduce an electrical signal to the distal ends causing rapid physiological changes within axons. This effect combined with the local changes in postsy-naptic signaling in dendrites may promote autophagosome biogenesis [37, 40].

2.2. Microtubule-dependent dynamics of neuronal autophagy

Previous data indicate that autophagy is a microtubule-dependent process. In cultured sympathetic neurons, autophagosomes formed in the distal ends of axon undergo retrograde transport along microtubules to the cell body where lysosomes that are necessary for the degradation step of autophagy are usually located [21]. Consistent with these observations, prominent retrograde transport of GFP-LC3-labelled autophagosomes has been observed in the axons of primary cerebellar granule cells [36]. In serum-deprived PC12 cells, autolysosomes formed by fusion of autophagosomes with lysosomes move in both anterograde and retrograde directions in neurites, and this trafficking requires microtubules [41]. Furthermore, both pharmacological and siRNA-based inhibition of directional microtubule motor proteins kinesin and dynein partially block respectively, anterograde and retrograde neuritic transport of autophagosomes, indicating that they participate in this transport. Recent observations in primary dorsal root ganglionic neurons support a maturation model in which autophagosomes initiate distally, engulfing mitochondria and ubiquitinated cargo, and move bidirectionally along microtubules driven by bound anterograde kinesin and retrograde dynein motors [22]. Fusion with late endosomes or lysosomes may then allow autophagosomes to escape from the early distal pool by robust retrograde dynein-driven motility. The involvement of the dynein-dynactin complex in the movement of autophagosomes along microtubules to lysosomes has also been demonstrated in non-neuronal cells [42]. Consistent with the formation of an autolysosomal compartment, autophagosomes increasingly acidify as they approach the cell soma, thereby fueling the catalysis of the degradation of their engulfed contents. Fully acidified autolysosomes undergo bidirectional motility suggesting reactivation of kinesin motors [22, 41].

The interaction of the autophagic membrane marker Atg8/LC3 with the microtubule-associated protein 1B (MAP1B) [43] implicates microtubule-dependent, axon-specific regulation of autophagosomes. Overexpression of MAP1B in non-neuronal cells reduces the number of LC3-associated autophagosomes without impairing autophagic degradation. The scarcity of LC3-labelled autophagosomes in CNS neurons under normal conditions may be explained by their high expression levels of MAP1B [10, 36]. By modifying microtubule function, the LC3-MAP1B interaction has been proposed to accelerate the delivery of LC3-autophagosomes to lysosomes, thereby promoting efficient autophagic turnover [37]. The exact mechanism underlying the involvement of microtubule in autophagosome formation, as well as targeting and fusion with lysosomes is open to debate [44, 45]. Based on (i) the absence of obvious changes in LC3 autophagosomes when they are associated with phosphorylated MAP1B-P, (ii) the elevated level of MAP1B-P bound to LC3 in dystrophic terminals containing a large number of autophagosomes [36] and, (iii) the conserved role of MAP1B-P in axonal growth and repair during development or injury (which implicates autophagy in remodelling axonal terminals during regeneration) [46], the interactions of LC3 with MAP1B and MAP1B-P have been proposed to represent a regulatory determinant of autophagy in axons under normal and pathological conditions respectively [37].

2.3. Functions of neuronal autophagy

Neurons, as non-dividing cells, are more sensitive to toxic components than dividing cells. Therefore, their survival and the maintenance of their specialized functions under physiological and pathological conditions is crucial requiring a tight quality control of cytoplasmic components and their degradation. Autophagy is believed to be of particular importance in the synaptic compartments of neurons where high energy requirements and protein turnover are necessary to sustain synaptic growth and activity. The CNS displays relatively low levels of autophagosomes under normal conditions, even after starvation, but requires an indispensable turnover of cytosolic contents by autophagy even in the absence of any disease-associated mutant proteins [10, 47, 48]. The scarcity of immature autophagosomes in neurons is likely to reflect a highly efficient autophagic degradation in the healthy brain. Accordingly, inhibition of autophagy causes neurodegeneration in mature neurons suggesting that autophagy may regulate neuronal homeostasis [11, 12]. For example, abnormal protein accumulation and eventual neurodegeneration are observed in the CNS of mice lacking the *atg5* or the *atg7* genes. This implies that basal autophagy is normally highly active and required for neuronal survival [11, 12]. The cardinal importance of autophagy in central neurons is further supported by recent studies showing a rapid accumulation of autophagosomes in cortical neurons when lysosomal degradation is inhibited. Thus, constitutive autophagy apparently plays an active role in neurons even under nutrient-rich conditions [49, 50].

2.3.1. Axonal homeostasis

Constitutive autophagy is probably essential for axonal homeostasis. Suppression of basal autophagy by either deleting an *atg* gene or inhibiting autophagic clearance in neurons disrupts axonal transport of vesicles destined for lysosomal degradation, and causes axonal swelling and dystrophy [11, 12, 37, 50]. For examples, Atg1/Unc-51 mutants in *C. elegans* show defaults in axonal membranes [51], and Unc-51.1, the murine homologue of Unc-51 is necessary for axonal extension, suggesting a possible role for these proteins in axonal membrane homeostasis [20, 52, 53]. In the cerebellum, neuron-specific deletion of FIP200, a protein implicated in autophagosome biogenesis, causes axon degeneration and neuronal death [13]. Altogether, these data suggest that autophagy is essential to maintain axonal structure and function through retrograde axonal transport [16]. The degree of vulnerability and the formation of intracellular inclusions vary significantly among the different types of CNS neurons in mutant brains deficient in Atg5 or Atg7 suggesting disparate intrinsic requirements for autophagy and relative levels of basal autophagy [20]. For example, while ubiquitinated inclusions are rare in the Atg5- or Atg7-deficient PCs, these cells are among the most susceptible neurons to *Atg 5/7* gene deletion [54, 55]. ULK1, the human homologue of Atg1 is incorporated into the active NGF-TrkA complex after its K-63 polyubiquitination and association with p62 [52, 56]. The subsequent interaction of ULK1 with endocytosis regulators allows trafficking of NGF-bound TrkA receptors into endocytic vesicles [57] providing a possible mechanism of crosstalk between autophagy and endocytosis. By fusing with autophagosomes, some membrane compartments, including endosomes, can be removed from axons and degraded in lysosomes. This process maintains the homeostasis of the axonal membrane networks and as such is essential for axonal physiology [20, 53].

Indeed, dysfunctional autophagy has been implicated in axonal dystrophy. Axonal swellings occur in autophagy-deficient mouse brains [11, 12] and genetic ablation of *Atg7* provokes cell-autonomous axonal dystrophy and degeneration, inferring that autophagy is crucial for membrane trafficking and turnover in axons [53]. In Atg5- or Atg7-deficient PCs, axonal endings exhibit an accumulation of abnormal organelles and membranous profiles much earlier than the somato-dendritic compartment [53, 54]. Axonal degeneration is increasingly believed to precede somatic death by a non-apoptotic auto-destructive mechanism [58, 59]. The "dying-back" progressive retrograde degeneration of the distal axon is a likely model of the chronic injury observed in neurodegenerative diseases [59]. NGF-deprivation induces autophagosome accumulation in the distal tips of neurites of PC12 cells, and knocking down *Atg7* or *beclin 1* expression delays neurite degeneration of NGF-deprived sympathetic neurons [60]. This suggests that overactive or deficient autophagy contributes to axonal degeneration in a dying-back manner due to the fragility of the axonal tips [20].

2.3.2. Dendritic autophagy

Early autophagosomes have also been observed in dendrites and the cell body of neurons suggesting that axon terminals are not be the only sites where neuronal autophagosomes form, and that autophagy may play a regulatory function in dendrites under physiological and pathological conditions [19]. Along this line, mTOR a key regulator of the autophagic pathway, modulates postsynaptic long-term potentiation and depression, suggesting that autophagy may critically control synaptic plasticity at the postsynaptic, dendritic compartment [61]. Further investigations are required to determine the specific roles of autophagy in dendrites and axons.

Since autophagosomes can fuse with endosomes and form amphisomes, there is a link between autophagy and endocytosis [62]. ESCRT proteins have recently been implicated in normal autophagy [19, 63, 64]. The endocytic pathways, in particular multi-vesicular bodies (MVBs) may serve as critical routes for autophagosomes to reach lysosomes, because defects in ESCRT function prevents fusion or maturation of autophagosomes. The ESCRT-MVB pathway could represent the primary, if not the only, route for delivering autophagosomes to lysosomes in some cell types [20]. In neurons, a large part of the endocytosed cargo merges with the autophagic pathway prior to being degraded by lysosomes [65]. Alterations in ESCRT function have also been linked to autophagy-deficiency in fronto-temporal dementia (FTD) and amyotrophic lateral sclerosis (ALS). In these cases, the particular vulnerability of the neurons appears to be associated with a dysfunction in the autophagosome-MVB pathway in the dendritic compartment [19].

2.3.3. Protein homeostasis

Neurons deficient in Atg5 or Atg7 exhibit an accumulation of polyubiquitinated proteins in inclusion bodies even though the proteasome function is normal, suggesting that basal autophagy prevents spontaneous protein aggregation and plays an essential role in protein clearance and homeostasis in neurons. Such a function is even more critical in neurons expressing disease-related proteins like the aggregate-prone mutant α-synuclein and poly-glutamine-containing proteins [66-68], although how autophagy selectively degrades these disease-related proteins is unclear. The ubiquitin-associated protein p62 is a likely candidate, providing a link between autophagy and selective protein degradation. Indeed, p62 binds numerous proteins through multiple protein-protein interacting motifs, including one for LC3 [55, 56] and the ubiquitin-associated C-terminal domain which binds ubiquitinated proteins. The relationship between p62 and autophagy is further supported by the observation that a marked accumulation of p62 and LC3 occurs only when lysosomes, but not proteasomes, are blocked. Furthermore, p62 protein levels are elevated in autophagy-deficient neurons [36, 55].

This argues that p62 is a specific substrate of autophagic degradation rather than a molecule involved in autophagosome formation since p62-knockout mice display intact autophagosomes and slower protein degradation. Autophagy-deficient cells and neurons accumulate ubiquitin- and p62-positive inclusions, and this accumulation is greatly reduced by ablating p62 [55, 69]. p62 with mutations in the LC3 recognition sequence escape autophagic degradation, leading to the formation of inclusions, whereas those with mutations in the self-oligomerizing domain PB1 are poorly degraded, but no protein inclusions form. Thus increased levels and oligomerization of p62 are required for the formation of inclusion bodies, and their degradation is facilitated by oligomerization. Ubiquitinated aggregates induced by proteasome inhibition are also reduced in p62-deficient cells suggesting that p62 is a general mediator of inclusion formation and normally functions as an adaptator targeting proteins for autophagic degradation [20].

2.3.4. Neuronal autophagy in synapse development, function and remodeling

Neuronal autophagy has been recently shown to play an important role in synapse development. The ubiquitin-proteasome system negatively regulates growth of the neuromuscular junction (NMJ) in *Drosophila* [70] whereas NMJ is positively regulated by neuronal autophagy; a decrease or an increase in autophagy correspondingly affects synapse size [71]. Indeed, an overexpression or a mutation of *Atg1*, a gene involved in autophagy induction, respectively enhanced or decreased NMJ growth. Furthermore, this positive effect of autophagy on NMJ development is mediated by downregulating Hiw, an E3 ubiquitin ligase which negatively regulates synaptic growth by downregulating Wallenda (Wnd), a MAP kinase kinase [70- 72]. Although autophagy is considered as a nonselective bulk degradation process, it can regulate specific developmental events in a substrate-selective mode [73, 74]. In *C. elegans*, when presynaptic afferences are removed from postsynaptic cells, GABAA receptors are selectively targeted to autophagosomes [73]. Accordingly, Hiw could traffic to autophagosomes via a still unknown mechanism, although Hiw could be unselectively degraded by autophagy along with other presynaptic proteins. Interestingly, the synaptic density in mice carrying an atg1 mutation is decreased due to excessive activity of the MAP kinase ERK, suggesting that activated ERK negatively regulates synapse formation and that Atg1 regulates synaptic structure by downregulating ERK activity [75]. As pointed out by Shen and Ganetzky [71], autophagy is a perfect candidate to modulate synaptic growth and plasticity in function of environmental conditions, resulting in plausible consequences in learning and memory.

Autophagy has recently been shown to regulate neurotransmission at the presynaptic level [76]. Besides enhancing protein synthesis via the mTORC1 complex, mTOR activity inhibits autophagy by an Atg13 phosphorylation-induced blockade of Atg1 [77]. In the nervous system, mTORC1 promotes learning and synaptic plasticity dependent on protein synthesis [78- 80]. Conversely, the mTOR inhibitor rapamycin impedes protein synthesis and blocks cell injury-induced axonal hyperexcitability and synaptic plasticity, as well as learning and memory [81, 82]. In prejunctional dopaminergic axons, inhibition of mTOR induces autophagy as shown by an increase in autophagosome formation, and decreases axonal volume, synaptic vesicle number and evoked dopamine release. Similarly, non-dopaminergic striatal terminals also display more autophagosomes and fewer synaptic vesicles. Conversely, chronic autophagy deficiency in dopamine neurons increases dopaminergic axon size and evoked dopamine release, and promotes rapid presynaptic recovery. Thus mTOR-dependent axonal autophagy locally regulates presynaptic structure and function. In cultured brain slices, the occurrence of autophagosomes in presynaptic terminals isolated from their cell bodies confirms that autophagosomes are locally synthesized [83], and supports the view that this autophagy may serve to modulate presynaptic terminal function by sequestrating presynaptic components [76]. The global stimulating effect of chronic autophagy deficiency on dopaminergic neurons is consistent with the implication of autophagy in neurite retraction of sympathetic neurons *in vitro* [84] and neuritic growth in developing neurons [21].

There are only a few other reports indicating that autophagy may participate in synapse remodeling. In the cerebellar cortex of the ($Bax^{-/-};GluR\delta2^{Lc}$) double mutant mouse (Fig. 2A), prominent autophagic profiles are evident in parallel fiber terminals subjected to intense remodeling in the absence of the PCs, their homologous target neurons [85]. As mentioned above, endocytosed GABAA receptors are selectively targeted to autophagosomes *in C. elegans* neurons [73], whereas autophagy promotes synapse outgrowth in *Drosophila* [71]. Autophagy may also modulate synaptic plasticity as recently demonstrated in mammalian hippocampal neurons [61]. Here, neuronal stimulation by chemical LTD induces NMDAR-dependent autophagy by inhibiting the PI3K-Akt-mTOR pathway. Enhanced autophagosome formation in the dendrites and spines of these neurons targets internalized AMPA receptors to lysosomes suggesting that autophagy contributes to the NMDAR-dependent synaptic plasticity required to maintain LTD and assure certain brain functions [61]. A possible mechanism for this formation of autophagosomes and autophagic degradation of AMPARs in dendritic shafts and spines may involve a change in endosome cycling. The formation of more amphisomes due to the fusion of endosomes with autophagosome [86, 87] would reduce the recycling endosome population, and direct more AMPAR-containing endosomes to autophagosomes for lysosomal degradation. Another alternative, but not exclusive actor is p62. This autophagosomal protein is important for LTP and spatial memory [88], interacts with AMPAR and is required for the trafficking of AMPAR [89]. AMPAR via its interaction with p62 would be trapped in autophagosomes as their number increased [61]. mTOR regulates protein turnover in neurons by functioning at the intersection between protein synthesis and degradation. During learning and reactivation in the amygdale and hippocampus, rapamycin inhibition of mTOR has recently been shown to impair object recognition memory [90], implicating signaling mechanisms involved in protein synthesis, synaptic plasticity and cell metabolism in this cognitive function.

2.4. Few autophagosomes, a feature of basal neuronal autophagy

Neurons are highly resistant to large-scale induction of autophagy in response to starvation, probably due to the multiple energy sources available to assure their function [48]. Interestingly, the activity of mTOR, a negative regulator of autophagy is significantly reduced in hypothalamic neurons from mice after a 48h starvation [91], although there are reports that autophagy in neurons can be regulated independently of mTOR [92, 93]. For example, insulin impairs the induction of neuronal autophagy *in vitro*, but in its absence induction of autophagy is mTOR-dependent. Furthermore, a potent Akt inhibitor provokes robust autophagy [92]. Thus insulin signaling maintains a low level of autophagosome biogenesis in healthy neurons constituting a critical mechanism for controlling basal autophagy in neurons. In addition to insulin signaling, multiple parallel signaling pathways including the mTOR pathway can regulate autophagy in neurons. From these data, Yue and collaborators [20] have proposed that basal autophagy in CNS neurons is regulated by at least two mechanisms: (1) a non-cell-autonomous mechanism whereby regulators (nutrients, hormones and growth factors) are supplied by extrinsic sources (glia, peripheral organs), (2) a cell-autonomous mechanism controlled by intrinsic nutrient-mediated signaling or specific factors expressed in neurons.

Neurons may depend less on autophagy to provide free amino acids and energy under physiological conditions given their quasi exclusive use of blood-born glucose as a source of carbon and energy for protein synthesis. Accordingly, the primary function of neuronal autophagy may be different than a primary response to starvation, and autophagy regulatory mechanisms are likely to be specific in neurons. Furthermore, gender differences in autophagic capacity have been suggested by the faster autophagic response to starvation of cultured neurons from male rats compared to those from females [94]. While *in vivo* evidence of neuronal autophagy mediated by nutrient signaling is still missing, a number of stress-related signals, neuron injuries and neuropathogenic conditions trigger prominent formation and accumulation of autophagosomes in neurons. During this process, neurons may undergo a significant change in autophagy regulation, involving a deregulation that allows neurons to switch from basal level (neuron-specific process featured by a low number of autophagosomes) to an activated state (well-conserved induced autophagy with large-scale biosynthesis of autophagosomes) [20]. Hypoxic-ischemia [95, 96], excitotoxicity [97-99], the dopaminergic toxins, methamphetamine and MPP+ [65, 100, 101], proteasome inhibition [102-104], lysosomal enzyme/lipid storage deficiencies [105-108] are examples of these pathological inducers of neuronal autophagy (see below).

3. Autophagy in neuronal physiopathology

Autophagy normally protects effect against neurodegeneration, but defects in the autophagy machinery are sufficient to induce neurodegeneration. Indeed, neuron-specific disruption of autophagy results in neurodegeneration [11, 12]; for example PC-specific Atg7 deficiency impedes axonal autophagy via an important p62-independent axonopathic mechanism associated with neurodegeneration [55]. Furthermore, specific defects in selective autophagic components or in the cargo selection process can induce neurodegeneration. This hypothesis is supported by the studies of cargo recognition and degradation components, such as p62, NBR1, or ALFY [109, 110]. Defects at any one of the autophagic steps can cause an abnormal accumulation of cytosolic components and lead to disease states. Therefore, each step of the autophagic process needs to be tightly regulated for efficient autophagic degradation.

The housekeeping role of neuronal autophagy is more evident when neurons are loaded with pathogenic proteins [67]. In many neurodegenerative disorders, cytoplasmic, nuclear and extracellular inclusions composed of aggregated and ubiquitinated proteins are believed to contribute to organelle damage, synaptic dysfunction and neuronal degeneration. The autophagic process in diseased neurons participates in the clearance of abnormal aggregate-prone proteins such as the expanded glutamine (polyQ)-containing proteins (e.g. mutant huntingtin in Huntington's disease (HD)), mutant forms of α-synuclein in familial Parkinson disease (PD), different forms of tau in Alzheimer's disease (AD), tauopathies and FTD, mutant forms of SOD1 in motor diseases such as ALS, and mutant forms of PMP22 in peripheral neuropathies are cleared from diseased neurons by autophagy [19, 20, 55, 56, 66, 67, 111-115]. However, accumulation of these intracellular aggregates is believed to play a significant role in the etiology of neurodegenerative diseases including prion diseases (PrD) [3, 67]. One common feature is the dramatic cyto-pathological accumulation of autophagosomes in injured and degenerating neurons [116-121]. Such signs of defects in autophagy have been interpreted as a result of an "autophagic stress", or in other words an imbalance between protein synthesis and degradation [116]. This has traditionally been viewed as a highly destructive cellular mechanism, driving the cell to death [117]. In these diseases, it is now accepted that autophagy eliminates aggregate-prone proteins and damaged organelles more efficiently than the proteasome machinery. Since the proteasome is unable to degrade them [122], the clearance of misfolded, aggregated proteins originating from neuropathologic deficits is highly dependent on autophagy. However, a blockade of the autophagic flux is likely to impede the clearance of these proteins. The accumulation of aggregated proteins and organelles within the diseased neurons then contributes to cell dysfunction and in the end results in cell death [16], (Fig. 3). Indeed, pharmacological upregulation of autophagy reduces neuronal aggregates and slows down the progression of neurological symptoms in animal models of tauopathy and HD [123], AD [41, 124, 125] and PrD [126, 127].

The mechanisms that determine the activation of autophagy for the removal of aggregated proteins are not clearly understood, but failure of the other proteolytic systems to handle the altered proteins seems to at least partly underlie autophagy activation. Thus oligomers and fibers of particular proteins can block the proteolytic activity of the ubiquitin-proteasome system and chaperone-mediated autophagy (CMA) that results in autophagy upregulation [128, 129]. In addition, sequestration of negative regulators of autophagy in the protein aggregates could also provoke activation of this pathway. Thus it has been shown that blockage of autophagy in neurons leads to the accumulation of aggregated proteins and neurodegeneration even in the absence of aggregate-prone proteins [11, 12]. Although the specific reasons for the failure of the proteolytic systems are unknown, factors such as enhanced oxidative stress and aging seem to precipitate entry into a late failure stage when the activity of all degradation systems are blocked or decreased, leading to accumulation of autophagic vacuoles and aggregates and finally cell death [130].

Autophagy protects against cell death in the case of growth factor withdrawal, starvation and neurodegeneration, but it is required for some types of autophagic cell death [131-134]. However, the role of autophagy as a positive mediator of cell death is not well understood in mammalian systems, although many studies suggest that impaired autophagy sensitizes cells and organisms to toxic insults. Atg1-dependent autophagy restricts cell growth [135]. Cells

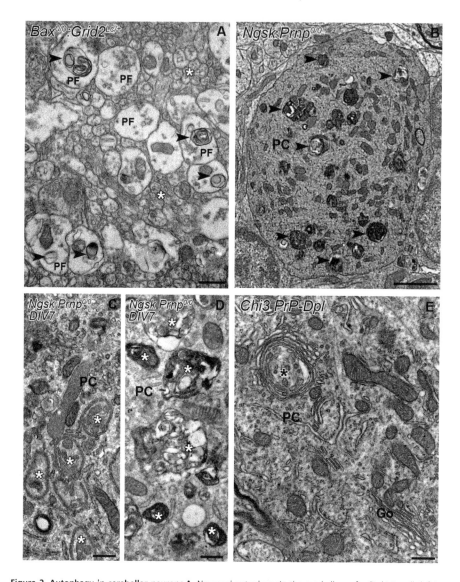

Figure 2. Autophagy in cerebellar neurons.A. Neuronal autophagy in the cerebellum of a Purkinje cell-deficient *Bax*$^{0/0}$*;Grid2*$^{Lc/+}$ double mutant mouse. Autophagic-like profiles (arrowheads) in presynaptic parallel fiber boutons (PF) in the cerebellar molecular layer. * intervaricose parallel fibers. **B.** Autophagolysosomes (arrowheads) characteristic of neuronal autophagy in the soma of a cerebellar Purkinje cell (PC) of a prion protein-deficient Ngsk *Prnp*$^{0/0}$ mouse. **C-D.** Phagophores, autophagosomes (* in C) and autophagolysosomes (* in D) in the soma of cerebellar Purkinje cells (PC) from prion protein-deficient Ngsk *Prnp*$^{0/0}$ mouse maintained 7 days *in vitro* (DIV7) in organotypic culture. **E.** Autophagosome (*) forming from a Golgi dictyosome in the Purkinje cell soma (PC) of a transgenic mouse expressing a neurotoxic Chi3 PrP-Dpl chimera. Go, normal Golgi dictyosome. Scale bars = 500 nm in A, C-E, 2 μm in B.

deficient in Pdk1, a positive regulator of mTOR pathway [136], display autophagy and reduced growth. The increased growth capacity that results from disrupting autophagy may contribute to the tumorigenicity of cells mutant for tumor suppressors [38, 137, 138]. Overexpression of Atg1 leads to apoptotic cell death [135]. Cells undergoing autophagic cell death display signs of apoptosis [139], as do Atg1-null cells [135]. Thus, elevated levels of autophagy promote cell death and the role of autophagy in cell death is likely to be context-dependent.

Neuronal autophagy is currently believed to constitute a protective mechanism that slows the advance of neurodegenerative disorders, and that its inhibition is associated with neurodegeneration [130]. Substantial attention is currently being focused on the molecular mechanisms underlying the autophagic fight against neurodegeneration, the role of autophagy in early stages of pathogenesis and therapeutical approaches to upregulate protective neuronal autophagy. It is unclear whether accumulation of autophagic vacuoles in degenerating neurons results from increased autophagic flux or impaired flux. A chronic imbalance between autophagosome formation and degradation causes "autophagic stress" [140]. Due to obvious therapeutic consequences, it is imperative to understand how autophagic stress occurs in each autophagy-associated neurodegenerative condition: either a cellular incapacity to support an excessive autophagic demand or a defective degradation (lysosomal) step [141].

4. Autophagy in prion diseases

4.1. Prion diseases

4.1.1. Infectious and familial prion diseases

Prion diseases (PrD) are transmissible spongiform encephalopathies (TSEs) which are fatal neurodegenerative diseases in humans (Creutzfeldt-Jakob disease (CJD), Gerstmann-Sträussler-Scheinker syndrome (GSS), variant CJD (vCJD), fatal familial insomnia (FFI) and kuru) and in animals (bovine spongiform encephalopathy (BSE), transmissible mink encephalopathy (TME), chronic wasting disease (CWD) and scrapie). In humans, PrD manifest after a long incubation period free of symptoms as a rapid progressive dementia that leads inevitably to death. Severe loss of neurons with extensive astrogliosis and moderate microglial activation, characteristic of all TSEs, results in a progressive spongiform degeneration of the brain tissue which is reflected by ataxia, behavioral changes and, in humans, a progressive cognitive decline [142-145]. According to the protein-only hypothesis [146], TSEs are caused by prions that are believed to be proteinaceous infectious particles mainly consisting of PrP^{SC}, an abnormal isoform of the normal, host-encoded prion protein (PrP^C), [142]. Prions are able to catalyze a switch from PrP^C conformation into an aggregated misfolded conformer PrP^{SC} which collects throughout the brain according to a prion strain-specific anatomo-pathologic signature. These PrDs share a protein misfolding feature with other neurodegenerative diseases (e.g. AD, PD and HD), [147].

The central role played by PrP^C in the development of PrD was first illustrated by the observation that disruption of the PrP gene (PRNP) in mice confers resistance to PrD and impairs

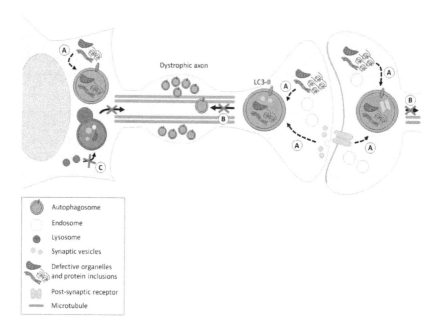

Figure 3. Impaired steps of neuronal autophagy in neurodegenerative disorders.A. Defective autophagosome biogenesis. B. Blockage of retrograde transport and accumulation of autophagosomes in dystrophic neurites. C. Failure of autophagosomes to fuse with lysosomes.

the propagation of infectious prions [148], while PrP-overexpressing (*tga20*) mice exhibit reduced incubation periods when compared with wild-type mice [149]. Overall, the current data argue for a primary role of the neuronal, GPI-anchored PrP^C in prion neuropathogenesis [150].The subversion of PrP^C function(s) as a result of its conversion into PrP^{SC} is assumed to account for prion-associated toxicity in neurons [151]. Whether PrP^{SC} triggers a loss of PrP^C physiological function (loss-of-function hypothesis) or promotes a gain of toxic activity (gain-of-function hypothesis), or both, is an ongoing debate in the TSE field [152]. Elucidating the roles of PrP^C in neurons should help to answer this question. Knockout experiments, however, have failed to reveal any obvious physiological role for PrP^C. Mice devoid of PrP^C are viable and display only minor phenotypic or behavioural alterations that vary according to the null strain, and hence, these results do not permit one to assign a specific function to PrP^C. *Ex vivo* studies support the involvement of PrP^C in copper homeostasis [153]. In addition, the localisation of PrP^C on the cell membrane and its affinity for the neuronal cell adhesion molecule (N-CAM), laminin, and the laminin receptor [154, 155] have implicated PrP^C in cell adhesion. Such properties may reflect the involvement of PrP^C in the outgrowth and maintenance of neurites, and even cell survival. Indeed, recent experimental evidence showing that PrP^C interacting with β1 integrin controls focal adhesion and turnover of actin microfilaments in neurons substantiates a role for PrP^C in neuritogenesis. Of note, integrins are well known inducers of autophagy (see review in chapter by Nollet and Miranti). Remarkably, during neuronal differentiation, the downregulation of Rho kinase (ROCK) activity by PrP^C is necessary for neurite sprouting [156]. A stress-protective activity has also been assigned to PrP^C based on results obtained with primary neuronal cultures. Neuronal cells derived from PrP-knockout mice are more sensitive to oxidative stress and serum deprivation than wild type cells [157-159]. Moreover, after ischemic brain injury, PrP^C-depleted mice revealed enlarged infarct volumes [160-162]. This neuroprotective role of PrP^C has been linked to cell signaling events. The interaction of PrP^C with the stress inducible protein (STI-1) generates neuroprotective signals that rescue cells from apoptosis [163]. Previous studies of both neuronal and non-neuronal cells substantiate the coupling of PrP^C to signaling effectors involved in cell survival, redox equilibrium and homeostasis (e.g. ERK1/2, NADPH oxidase [164], cyclic AMP-responsive element binding protein (CREB) transcription factor and metalloproteinases [165, 166]. According to these data, PrP^C has been proposed to function as a « dynamic cell surface platform for the assembly of signaling modules » [167]. Despite these overall advances, the sequence of cellular and molecular events that leads to neuronal cell demise in TSEs remains obscure [168, 169]. At present, one envisions that neuronal cell death results from several parallel, interacting or sequential pathways involving protein processing and proteasome dysfunction [170], oxidative stress [159, 171], apoptosis and autophagy [172].

4.1.2. Autophagy in prion-infected neurons

Prion propagation involves the endocytic pathway, specifically the endosomal and lysosomal compartments that are implicated in trafficking and recycling, as well as the final degradation of prions. Shifting the equilibrium between propagation and lysosomal clearance impairs the cellular prion load. This and the presence of autophagic vacuoles in prion diseased neurons [173, 174] suggest a role for autophagy in prion infection (reviewed in [172]). Indeed, the high numbers of autophagic vacuoles observed in neurons from experimentally prion-infected mice (Fig. 4A, B, Fig. 5) and hamsters is indicative of a robust activation of autophagy [175, 176]. Furthermore, autophagic vacuoles and multivesicular bodies have been detected in prion-infected neuronal cells in vitro [177].The formation of autophagic vacuoles has recently been observed in neuronal pericarya, neurites and synapses of neurons experimentally infected with scrapie, CJD and GSS [174], as well as in neuronal synaptic compartments in humans with certain PrD [173]. PrDs are further correlated with autophagy given that the Scrg1 protein (encoded by the scrapie responsive gene1, *Scrg1*) is upregulated in scrapie and BSE-infected brains, as well as in brains of patients with sporadic CJD [178-180] and is associated with neuronal autophagosomes [181, 182]. This Scrg1 protein is thus, a new marker for autophagic vacuoles in prion-infected neurons (Fig. 4A, B). In the brains of CJD and FFI patients and experimentally scrapie-infected hamsters, increased cytoplasmic levels of LC3-II-immunos-tained autophagosomes have been demonstrated in neurons, again indicating autophagy activation. In addition, the decreased p62 and polyubiquitinated proteins levels in hamster and human brains infected with prion suggest an upregulation of autophagy with enhanced autophagic flux and protein degradation. Downregulation of the mTOR pathway and upregulation of the beclin 1 pathway in these infected tissues provide further evidence of autophagy activation [183]. On the basis of these observations, Xu *et al.* [183] propose that neuronal autophagy is an intricate element of prion infections. They suggest that once PrP^{SC} enters host cells and is delivered to endosomes, it accumulates in amphisomes via fusion with autophagosomes and then with lysosomes. At this initial stage of infection, PrP^{SC} does not co-localize with autophagosomes, probably because PrP^{SC} levels are too low to be detected due to their rapid degradation in autophagolysosomes. In agreement with this explanation, blocking the fusion of autophagosomes with lysosomes using bafilomycin A1 permits the visualization of PrP-PG14 and PrP^{SC} in autophagosomes [183], as is the case for Aβ1-42 [184].

The role of lysosomes in PrDs is still controversial. Although autophagic lysosomal degradation of PrP^{SC} in infected neurons is supposed to clear prion aggregates and inhibit PrP^{SC} replication, there are indications that PrP^{SC} may subvert the autophagic-lysosomal system to promote the conversion of PrP^{C} into PrP^{SC}. Lysosomal inhibitors prevent the build-up of PrP^{SC} [126] and agonists of the autophagy-lysosome pathway enhance the clearance of PrP^{SC} [185, 186, 126]. However, as PrP^{SC} production increases, the accumulating PrP^{SC} may saturate the clearance capacity of the system causing lysosomal disruption and release of PrP^{SC} aggregates into the neuroplasm. In turn this would cause cell stress and over-activate autophagy, as has been reported in prion-diseased brain tissue [183].

The octapeptide repeats region of PrP^{C} has been shown to negatively influence autophagy. As measured by LC3-II expression, autophagy induced by serum deprivation occurs earlier and to a greater extent in hippocampal neurons from ZH-I *PrnP*$^{-/-}$ compared with those from wild type mice. Reintroduction of PrP^{C}, but not *PrP*^{C} lacking its N-t octapeptide region, into ZH-I PrnP$^{-/-}$neurons delays this upregulation of autophagy [187]. The transconformation of PrP^{C} into PrP^{SC} could interfere with the function of this domain and as a consequence, upregulate autophagy. It is conceivable that the activation of autophagy observed in PrD models reflects a defense mechanism designed to degrade prions and resist oxidative stress. A reduction in autophagy combined with endosomal/lysosomal dysfunction has indeed been proposed to contribute to the development of PrD [188]. Furthermore, the anti-cancer drug imatinib has been shown to activate lysosomal degradation of PrP^{SC} [186] and is a potent autophagy inducer [189]. When administered early during peripheral infection, imatinib delays both PrP^{SC} neuroinvasion and the onset of clinical disease in prion-infected mice [190]. Upregulation of autophagy has beneficial effects on the clearance of aggregate-prone proteins in PrD and other neurodegenerative diseases [66, 111-115, 191, 192]. Both lithium and trehalose enhance PrP^{SC} clearance from prion-infected cells by inducing autophagy, as demonstrated by increases in LC3-II protein and the number of GFP-LC3 puncta [193, 126]. Furthermore, PrP^{SC} can be cleared not only by mTOR-independent autophagy (lithium and trehalose), but also by the mTOR-dependent route because the mTOR inhibitor rapamycin also causes a decrease in cellular PrP^{SC}. Lithium-induced autophagy also reduces PrP^{C} levels. This treatment causes internalization of PrP^{C} [194], and the consequent reduction of available membrane-bound PrP^{C} is known to decrease its conversion into pathologic PrP^{SC} [195-198]. This would provide an additional, indirect way to reduce PrP^{SC} by reducing of PrP^{C} with lithium treatment.

Whether autophagy-inducing compounds are candidates for therapeutic approaches against prion infection has recently been investigated in prion-infected mice. Starting in the last third of the incubation periods, treatment with rapamycin and to a lesser extent with lithium significantly prolonged incubation times compared to mock-treated control mice [126, 172]. Along this line, activation of the class III histone deacetylase Sirtuin 1 (Sirt1) has been shown to mediate the neuroprotective effect of resveratrol against prion toxicity [199] and prevent prion protein-derived peptide 106-126 (PrP106-126) neurotoxicity via autophagy processing [200]. Moreover, Sirt1-induced autophagy protects against mitochondrial dysfunction induced by PrP106-126, whereas siRNA knockdown of Sirt1 sensitizes cells to PrP106-126-induced cell death and mitochondrial dysfunction. Finally, knockdown of Atg5 decreases LC3-II protein levels and blocks the effect of a Sirt1 activator against PrP106-126-induced mitochondrial dysfunction and neurotoxicity. Thus inducing Sirt1-mediated autophagy may be a principal neuroprotective mechanism against prion-induced mitochondrial apoptosis. Nevertheless, understanding the mechanisms underlying Sirt1-mediated autophagy against prion neurotoxicity and mitochondrial damage merits further investigation, in particular determining the Sirt1-mediated dowstream signaling network, including FOXOs, p53 and PGC-1α. More recently, the mTOR inhibitor and autophagy inducer rapamycin has been shown to delay disease onset and prevent PrP plaque deposition in a mouse model of the Gerstmann-Sträussler-Scheinker PrD [127]. Here, the reduction in symtom severity and prolonged survival correlate with increases in LC3-II levels in the brains of treated mice, suggesting that autophagy induction enhances elimination of misfolded PrP before plaques form. This is in agreement with the well known neuroprotective effects of rapamycin in various models of neurodegenerative diseases with misfolded aggregate-prone proteins (e.g. PD [111], ALS [201], HD [115], spinocerebellar ataxia [66, 202], FTD [203] and AD [41, 124, 125].

4.2. Doppel-expressing prion protein-deficient mice

Research efforts to determine the function of PrP^C using knockout mutant mice have revealed that large deletions in the PrP^C genome result in the ectopic neuronal expression of the prion-like protein Doppel (Dpl) causing late onset degeneration of PCs and ataxia in PrnP$^{-/-}$mouse lines, such as Ngsk [204], Rcm0 [205], ZH-II [206] and Rikn [207].

Similar PC degeneration is observed when the N-terminal truncated form of PrP is expressed (ΔPrP) in Prnp-ablated mouse lines [208] and when Dpl is overexpressed [209, 210]. Of note, full-length PrP^C antagonizes the neurotoxic effects of both Dpl and ΔPrP [208-212], but not PrP^C lacking the N-terminal residues 23-88 [213]. These results imply that Dpl and ΔPrP induce cell death by the same mechanism, likely by interfering with a cellular signaling pathway essential for cell survival and normally controlled by full-length PrP^C [209, 214]. The mechanism underlying Dpl-induced neurotoxicity is still under debate. PrP-deficient neurons undergo Dpl-induced apoptosis in a dose-dependent, cell autonomous manner [215]. Oxidative stress is a likely candidate to play a role in the death of these neurons because NOS activity is induced by Dpl both in vitro and in vivo [212, 216]. Endogenous, as well as exogenous PrP^C has been shown to inhibit Dpl-induced apoptosis, a neuroprotective function that has been attributed to its BCL-2-like properties [158]. Like BCL-2, PrP^C antagonizes mitochondrial apoptotic pathways, thereby protecting neurons from cell death [217- 219]. In BAX-induced apoptosis [220, 221], PrP^C probably acts by preventing the conformational changes in BAX that are necessary for its activation [222]. In primary cultures, Dpl-induced apoptosis of Prnp$^{+/+}$ as

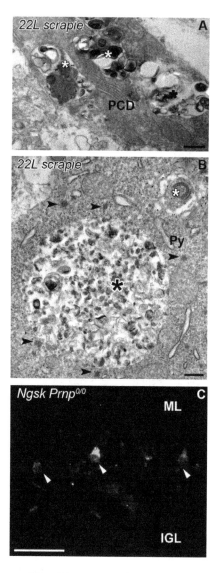

Figure 4. Scrapie responsive gene 1 (Scrg1)-immuno-cytochemistry in prion-infected and prion protein-deficient neurons. A-B. Scrg1 immunogold labeling in central neurons of a clinically ill 22L-scrapie-infected mouse. Scrg1-bound immunogold particles label autophagolysosomes (* in A) in a Purkinje cell dendrite (PCD) and an autophagosome forming from a Golgi dictyosome (white asterisk in B) in the somatic neuroplasm of a pyramidal neuron (Py) of the CA3 field of the hippocampus. In this neuron, lysosomes (arrowheads) and immunogold particles labelling Scrg1 surround a large autolysosome-like vacuole (black asterisk). **C.** Scrg1 immuno-fluorescent labeling of Purkinje cells (arrowheads) in the cerebellar cortex of a prion protein-deficient Ngsk $Prnp^{0/0}$ mouse. IGL, internal granular layer; ML, molecular layer. Scale bars = 500 nm in A-B and 50 μm in C.

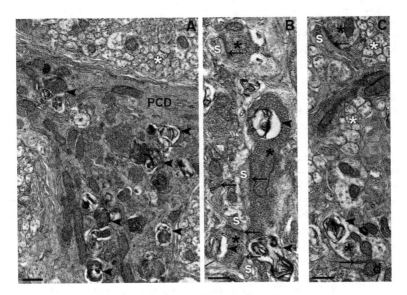

Figure 5. Neuronal autophagy in cerebellar neurons of a clinically ill 22L-scrapie-infected mouse. A. Accumulation of autophagosomes (arrowheads) in a main dendrite of a Purkinje cell (PCD) in the cerebellar molecular layer. *, parallel fibers. **B.** Autophagosomes (arrowheads) in presynaptic axon terminals (black asterisks) establishing synapses (arrows) on postsynaptic Purkinje cell dendritic spines (s). **C.** An intact parallel fiber bouton (black asterisk) makes a synapse (arrow) on a Purkinje cell spine (s) in the upper part of the picture and another parallel fiber bouton (black asterisk) containing an autophagosome (arrowhead) makes a synapse (arrow) on a putative interneuron dendrite (d) in the bottom of the picture. Scale bars = 500 nm.

well as *PrnP⁻/⁻* granule cells, has recently been shown to be inhibited by BAX deficiency or pharmacologically blocking caspase-3 suggesting that it is mediated by Bax and caspase-3 [223]. These results further confirm *in vivo* data concerning the effects of Bax expression on PC survival in the cerebellum of the Dpl-overexpressing Ngsk *PrnP⁻/⁻*mouse that we reported several years ago [224]. In these mice, PC death is already significant as early as 6 months of age. During aging, quantification of PC populations shows that significantly more PCs survived in the Ngsk *PrnP⁻/⁻:Bax⁻/⁻* double mutant mice than in the Ngsk *PrnP⁻/⁻*mice. However, the number of surviving PCs is still lower than wild type levels and less than the number of surviving PCs in *Bax⁻/⁻* mutants. This suggests that neuronal expression of Dpl activates both BAX-dependent and BAX-independent pathways of cell death. Interestingly, a partial rescue of Ngsk *PrnP⁻/⁻*PCs is observed in Ngsk *PrnP⁻/⁻-Hu-bcl-2* double mutant mice, in a proportion similar to that found in Ngsk *PrnP⁻/⁻:Bax⁻/⁻* mice, strongly supporting the involvement of BCL-2-dependent apoptosis in Dpl neurotoxicity [225]. The capacity of BCL-2 to apparently compensate for the deficit in PrP^C by partially rescuing PCs from Dpl-induced death suggests that the BCL-2-like property of PrP^C may counteract Dpl-like neurotoxic pathway in wild-type neurons. Although not exactly identical to BCL-2, PrP^C may functionally replace BCL-2 as it decreases in the aging brain [222]. The N-terminal domain of PrP^C which is partially homologous to the BH2 domain of BCL-2 family of proteins [226, 227] is probably responsible for the neuroprotective functions of PrP^C because BAX-induced apoptosis cannot be counteracted by N-terminally truncated PrP. BCL-2 antagonizes the pro-apoptotic effect of BAX by interacting directly with this BH2 domain [228-230], and this domain is missing in both Dpl and the neurotoxic mutated forms of PrP: ΔPrP [208, 214, 231] and Tg(PG14)PrP [232]. Interestingly, expression of Dpl fused to a BH2-containing octapeptide repeat and the N-terminal half of the hydrophobic region of PrP^C makes cells resistant to serum deprivation [233]. Furthermore, N-terminal deleted forms of PrP^C have been reported to activate both BAX-dependant and BAX-independant apoptotic pathways [231].

4.2.1. Autophagy in prion protein-deficient mice

The Dpl-activated, BAX-independent cell death mechanism may involve neuronal autophagy as we have detected the expression of Scrg1, a novel protein with a potential link to autophagy in the Ngsk *PrnP$^{-/-}$*PCs (Fig. 4C), [181]. Both neuronal Scrg1 mRNA and protein levels are increased in prion-diseased brains [179, 180], and Scrg1 is associated with dictyosomes of the Golgi apparatus and autophagic vacuoles in degenerating neurons of scrapie-infected Scrg1-overexpressing transgenic and WT mice (Fig. 4A, B), [181, 182]. Both before and during PC loss, protein levels of Scrg1 and the autophagic markers LC3-II and p62 are increased in Ngsk *PrnP$^{-/-}$*PCs, whereas their mRNA expression is stable, suggesting that the degradation of autophagic products is impaired in these neurons [234, 235]. Autophagic profiles collect in somato-dendritic and axonal compartments of Ngsk *PrnP$^{-/-}$* (Figs. 2B, 6), but not wild-type PCs. The most robust autophagy occurs in dystrophic profiles of the PC axons in the cerebellar cortex (Fig. 6D) and at their preterminal and terminal levels in the deep cerebellar nuclei (Fig. 6A-C) suggesting that it initiates in these axons. Taken together, these data indicate that Dpl triggers autophagy and apoptosis in Ngsk *PrnP$^{-/-}$* PCs. As reflected by the abundance of autophagosomes in the diseased Ngsk PCs, Dpl neurotoxicity induces a progressive dysfunction of autophagy, as well as apoptosis. Whether this autophagy dysfunction triggers apoptotic cascades or provokes autophagic cell death independent of apoptosis remains to be resolved. In the Ngsk *PrnP$^{-/-}$* PCs, the increased expression of LC3-II and p62 at the protein level, without any change in mRNA levels, suggests that the ultimate steps of autophagic degradation are impaired. This is further confirmed by the prominence of autophagolysosomes in these neurons which indicate that the fusion of autophagosomes with lysosomes occurs normally, but downstream, the autophagic flux is blocked.

To further investigate the neurodegenerative mechanisms induced by Dpl in Ngsk cerebellar PCs, we are using an organotypic cerebellar culture system which allows an easier way to approach mechanistic questions than in vivo models [236]. For this purpose, we have assessed the growth and viability of PCs in cerebellar organotypic cultures from Ngsk and ZH-I *PrnP$^{-/-}$* mice using morphometric methods to measure PC survival and development [237]. The timing and amplitude of PC growth impairment and neuronal death are similar in Ngsk and ZH-I *PrnP$^{-/-}$* cultures (Fig. 7). In addition, increased amounts of autophagic (LC3-II, Fig. 8) and apoptotic (caspase-3, Fig. 9) markers are detected in protein extracts from both cultures indicating that both apoptosis and autophagy (Fig. 2C, D) contribute to PC death in Ngsk [235] and ZH-I cultures. This suggests that PrP^C -deficiency, rather than Dpl expression, is respon-

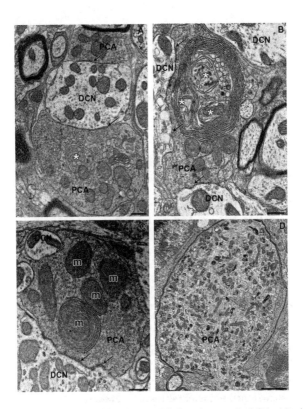

Figure 6. Neuronal autophagy in the cerebellar deep nuclei of a prion protein-deficient Ngsk *Prnp*⁰/⁰ mouse. A. A presynaptic terminal of a Purkinje cell axon (PCA) establishes symmetric synapses on a postsynaptic dendrite from a neuron of the interpositus deep cerebellar nucleus (DCN) and contains an autophagosome (*). **B.** A double membrane wrap sequesters autophagosomes (*) in a Purkinje cell axon varicosity (PCA) symmetrically synapsing (arrows) on dendrites from neurons of the dentate deep cerebellar nucleus (DCN). **C.** Mitophagy by double membranes wrapping around mitochondria (m) in a Purkinje cell presynaptic axon terminal (PCA) making symmetric synapses (arrows) on postsynaptic dendrites from dentate deep nuclear neurons (DCN). **D.** Dystrophic Purkinje cell axon (PCA) filled with electron-dense autophagic profiles in the cerebellar internal granular layer. Scale bars = 500 nm in A-C, 2 μm in D.

sible for the neuronal growth deficit and loss in these cultures. For presently unknown reasons, the neurotoxic properties of Dpl do not seem to contribute to the degeneration of Ngsk PCs in these organotypic cultures. As the neurotoxicity induced by Dpl takes about 6 months to develop *in vivo*, it is possible that organotypic cultures are not mature enough to model 6-month-old cerebellar tissue. Nevertheless, *ex vivo* cerebellar organotypic cultures do provide a suitable system for analyzing the mechanisms underlying the neurotoxic effects of PrP^C-deficiency and prion infections [238] using pharmacological and siRNA-based approaches.

Our results have shown that PrP^C has a neuroprotective role in cerebellar PCs. As PCs survive *in vivo* in the cerebellum of the ZH-I mouse, the death of the ZH-I PCs in the organotypic

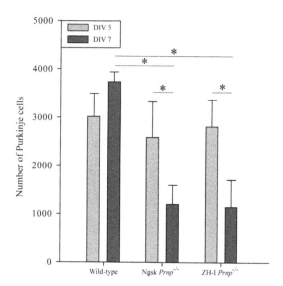

Figure 7. Purkinje cell loss in cerebellar organotypic cultures from wild-type and *Prnp*-deficient mice. PCs stained for calbindin by immuno-fluorescence were counted. This analysis reveals similar, significant reductions in the number of PCs between DIV5 and DIV7 for Ngsk *Prnp*[-/-] (53.5%) and ZH-I *Prnp*[-/-] (59%) cultures. During this period, the number of PCs in wild-type cultures is stable ($p > 0.05$). Although the number of PCs is not significantly different between genotypes at DIV5, by DIV7 there are similar decreases in mutant organotypic cultures (Ngsk: 67.8% and ZH-I: 69%) compared to wild-type cultures (two-way ANOVA followed by post-hoc Tukey test; * $p < 0.001$).

cultures is likely to stem from the inherent stress of the *ex vivo* conditions. As mentioned above, PrP^C negatively regulates autophagy as demonstrated by the upregulation of autophagy following serum deprivation in *PrnP*[-/-] hippocampal neurons when compared to PrP^C-expressing neurons [187]. Recent results suggest that PrP^C can directly modulate autophagic cell death. Using antisens oligonucleotides targeting the *Prnp* transcript, the downregulation of PrP^C expression in glial and non-glial tumor cells induces autophagy-dependent, apoptosis-independent cell death [239]. Previous data have shown that PrP^C acts as a SOD [240] and modulates the activity of Cu/Zn SOD by binding 5 Cu^{++} ions on its N-terminal octapeptide repeat domain [153, 157, 241]. A recent study of the effects of H$_2$O$_2$-induced oxidative stress on hippocampal neurons expressing PrP^C or deficient in PrP^C provides further support for the protective role of PrP^C against oxidative stress [242]. Although autophagy and apoptosis occur in both lines, the *Prnp*[-/-] neurons are less resistant to H$_2$O$_2$-induced oxidative stress than the *Prnp*[+/+] neurons confirming the anti-oxidant activity of PrP^C.

Furthermore, autophagy is more enhanced in *Prnp*[-/-] neurons than in *Prnp*[+/+] neurons. In the latter, this is due to H$_2$O$_2$-induced enhancement of autophagic flux, and in the former due to H$_2$O$_2$-induced impairment of autophagic flux. Similarly, experiments using Atg7 siRNA to inhibit autophagy have revealed that the increased autophagic flux in Prnp[+/+] neurons protects against H$_2$O$_2$ cytotoxicity. Thus a deficiency in Prnp may impair autophagic flux via H$_2$O$_2$-

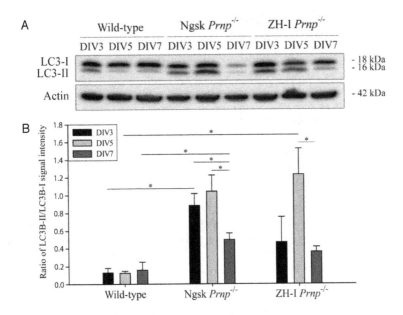

Figure 8. A. Western blot of the autophagic marker LC3B-II, in extracts prepared from organotypic cultures from wild-type, Ngsk and ZH-I *Prnp$^{-/-}$* mouse cerebellum at DIV3, 5 and 7. Actin was used as a loading control. **B.** Autophagy was measured by quantifying the ratio of the band intensities of LC3B-II and LC3B-I (n ≥ 3 mice) which reflects the amount of autophagosomes. Compared to wild-type cultures, this ratio increases in mutant cultures at DIV5 suggesting enhanced autophagy and then decreases at DIV7 probably as a result of either autophagic degradation or PC death (Kruskal-Wallis test followed by post-hoc Tukey test; * $p < 0.05$).

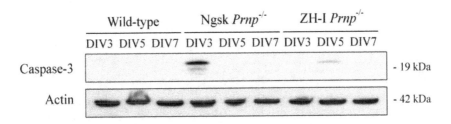

Figure 9. Western blot of the pro-apoptotic activated caspase-3. Activated caspase-3 is detected in extracts of organotypic cultures from Ngsk and ZH-I *Prnp$^{-/-}$* mouse cerebellum, but not from wild-type mouse cerebellum. Actin serves as a loading control.

induced oxidative stress contributing to autophagic cell death [242]. Since autophagic flux is apparently normal in both *Prnp$^{+/+}$* and *Prnp$^{-/-}$* neurons in the absence of stress, the lack of PrP^C only seems to contribute to autophagy impairment under stress-induced conditions, such as H$_2$O$_2$ treatment [242], stress-inducing *in vitro* conditions, as well as Dpl-induced toxicity.

4.2.2. Prion protein PrP-doppel Dpl chimeras

When overexpressed ectopically in neurons, mutations within the central region of PrP^C are associated with severe neurotoxic activity, similar to that of Dpl [231, 243]. The absence of these segments, called central domains (CD) is believed to be responsible for neurodegeneration and ataxia. To understand the dual neurotoxicity *vs.* neuroprotective roles of PrP^C, transgenic mice expressing a fusion protein made of the CD of PrP^C inserted within the Dpl sequence have been generated [244]. These mice failed to develop typical Dpl-mediated neurological disorder indicating that this N-terminal portion of PrP^C reduces Dpl toxicity. To further investigate Dpl-like neurotoxicity, Lemaire-Vieille *et al.* recently generated lines of transgenic mice expressing three different chimeric PrP-Dpl proteins [245]. Chi1 (Dpl 1-57 replaced with PrP 1-125) and Chi2 (Dpl 1-66 replaced with PrP1-134) abrogates the pathogenecity of Dpl confirming the neuroprotective role of the PrP 23-134 N-terminal domain against Dpl toxicity. However, when Dpl 1-24 were replaced with PrP 1-124, these Chi3 transgenic mice that express a very low level of the chimeric protein develop ataxia, as early as 5 weeks of age. This phenotype is only rescued by overexpressing PrP^C, and not by a single copy of full-length PrP^C, indicating the strong toxicity of the chimeric protein Chi3. The Chi3 mice exhibit severe cerebellar atrophy with significant granule cell loss and prominent signs of autophagy in PCs (Fig. 2E). We conclude that the first 33 amino acids of Dpl, that are absent in Chi1 and Chi2 constructs, confer toxicity to the protein. This is confirmed *in vitro* by the highly neurotoxic effect of the 25-57 Dpl peptide on mouse embryo cortical neurons. Since this chimeric transgene is not expressed by PCs in the transgenic mice expressing Chi3, the signs of autophagy displayed by these neurons in vitro could result from the neurotoxic effect of the exogenous Chi3 chimeric protein, as well as the deleterious effect of losing their primary afferences (i.e. the granule cells).

5. Perspectives

The beneficial effects that autophagy has on prion infections is currently supported by a growing bulk of evidence from *in vivo* and *ex vivo* data and is strongly promising for future mid-term therapeutic approaches. To further understand the fascinating interplay between autophagy and PrDs, further investigations are necessary to decipher their molecular interactions. Important issues remain. How are the different phases of prion infection physiopathology i.e. propagation, trafficking, recycling and clearance connected with autophagy? Which autophagic pathways are activated by prions - the mTOR-dependent, mTOR-independent or both? The biological function of autophagy per se in prion infection is obscure as the cellular levels of autophagy can apparently modify cell susceptibility to prion infection, although changes in autophagy may be a pre-requisite or a consequence of a prion infection.

Overall, the results point to a need to counteract cell stress and to eliminate toxic aggregate-prone proteins that eventually saturate the usual degradation pathways, including autophagy. These are common features of prion disesase and most of the other neurodegenerative diseases described in this review. Saturation of the autophagic machinery, loss or imbalance of autophagic flux is believed to lead to neurodegeneration. Understanding how autophagy relates to these diseases is a first step for developing autophagy modulation-based therapies for treating neurological disorders. This implies therapeutic consideration for each type of autophagic default at a precise step of the neurodegenerative disease concerned.

Acknowledgements

The authors are greatly indebted to Dr. Anne-Marie Haeberlé (CNRS UPR3212, Strasbourg) for excellent assistance in transmission electron microscopy and prion-infected tissue handling as well as to Dr. Catherine Vidal (Institut Pasteur, Paris) for intra-cerebellar inoculation of mice with 22L scrapie. A. R. is supported by a doctoral grant from the French Minister of Research and Technology and A. G. is supported by a grant from the French Centre National de la Recherche Scientifique and the AgroParisTech High School.

Author details

Audrey Ragagnin[1], Aurélie Guillemain[1], Nancy J. Grant[2] and Yannick J. R. Bailly[1]

1 Cytologie & Cytopathologie Neuronales, INCI CNRS UPR3212, Université de Strasbourg, Strasbourg, France

2 Trafic Membranaire Dans les Cellules Neurosécrétrices et Neuroimmunitaires, INCI CNRS UPR3212, Université de Strasbourg, Strasbourg, France

References

[1] Mizushima N, Komatsu M. Autophagy: renovation of cells and tissues. Cell (2011) 147:728-41.

[2] Levine B, Deretic V. Unveiling the roles of autophagy in innate and adaptive immunity. Nat Rev Immunol (2007) 7(10):767-77.

[3] Levine B, Kroemer G. Autophagy in the pathogenesis of disease. Cell (2008) 132:27-42.

[4] Mizushima N, Levine B, Cuervo AM, Klionsky DJ. Autophagy fights disease through cellular self-digestion. Nature (2008) 451:1069-75.

[5] Clarke PG. Developmental cell death: morphological diversity and multiple mechanisms. Anat Embryol (1990) 181:195-213.

[6] Kroemer G, Levine B. Autophagic cell death: the story of a misnomer. Nat Rev Mol Cell Biol (2008) 9:1004-10.

[7]   Novikoff AB, Beaufay H, De Duve C. Electron microscopy of lysosomerich fractions from rat liver. J Biophys Biochem Cytol (1956) 2:S179-84.

[8]   De Duve C. The significance of lysosomes in pathology and medicine. Proc Inst Med Chic (1966) 26:73-6.

[9]   Dixon JS. "Phagocytic" lysosomes in chromatolytic neurones. Nature (1967) 215:657-658.

[10] Mizushima N, Yamamoto A, Matsui M, Yoshimori T, Ohsumi Y. In vivo analysis of autophagy in response to nutrient starvation using transgenic mice expressing a fluorescent autophagosome marker. Mol Biol Cell (2004) 15:1101-11.

[11] Hara T, Nakamura K, Matsui M, Yamamoto A, Nakahara Y, Suzuki-Migishima R, Yokoyama M, Mishima K, Saito I, Okano H, Mizushima N. Suppression of basal autophagy in neural cells causes neurodegenerative disease in mice. Nature (2006) 441:885-9.

[12] Komatsu M, Waguri S, Chiba T, Murata S, Iwata J, Tanida I, Ueno T, Koike M, Uchiyama Y, Kominami E, Tanaka K. Loss of autophagy in the central nervous system causes neurodegeneration in mice. Nature (2006) 441:880-4.

[13] Liang CC, Wang C, Peng X, Gan B, Guan JL. Neural-specific deletion of FIP200 leads to cerebellar degeneration caused by increased neuronal death and axon degeneration. J Biol Chem (2010) 285:3499-509.

[14] Xie Z, Klionsky DJ. Autophagosome formation: core machinery and adaptations. Nat Cell Biol (2007) 9:1102-9.

[15] He C, Klionsky DJ. Regulation mechanisms and signaling pathways of autophagy. Annu Rev Genet (2009) 43:67-93.

[16] Lee J-A. Neuronal autophagy: a housekeeper or a fighter in neuronal cell survival? Exp Neurobiol (2012) 21:1-8.

[17] Klionsky D, et 1268 al. Guidelines for the use and interpretation of assays for monitoring autophagy. Autophagy (2012) 8:445-544.

[18] Mizushima N, Yoshimori T, Ohsumi Y. The role of Atg proteins in autophagosome formation. Annu Rev Cell Dev Biol (2011) 27:107-32.

[19] Lee JA, Beigneux A, Ahmad ST, Young SG, Gao FB. ESCRT-III dysfunction causes autophagosome accumulation and neurodegeneration. Curr Biol (2007) 17:1561-1567.

[20] Yue Z, Friedman L, Komatsu M, Tanaka K. The cellular pathways of neuronal autophagy and their implication in neurodegenerative diseases. Biochim Biophys Acta (2009) 1793:1496-507.

[21] Hollenbeck PJ. Products of endocytosis and autophagy are retrieved from axons by regulated retrograde organelle transport. J Cell Biol (1993) 121:305-15.

[22] Maday S, Wallace KE, Holzbaur EL. Autophagosomes initiate distally and mature during transport toward the cell soma in primary neurons. J Cell Biol (2012) 196:407-17.

[23] Dunn WA Jr. Studies on the mechanisms of autophagy: formation of the autophagic vacuole. J Cell Biol (1990) 110:1923-33.

[24] Axe EL, Walker SA, Manifava M, Chandra P, Roderick HL, Habermann A, Griffiths G, Ktistakis NT. Autophagosome formation from membrane compartments enriched in phosphatidylinositol 3-phosphate and dynamically connected to the endoplasmic reticulum. J Cell Biol (2008) 182:685-701.

[25] Hayashi-Nishino M, Fujita N, Noda T, Yamaguchi A, Yoshimori T, Yamamoto A. A subdomain of the endoplasmic reticulum forms a cradle for autophagosome formation. Nat Cell Biol (2009) 11:1433-7.

[26] Ylä-Anttila P, Vihinen H, Jokitalo E, Eskelinen EL. 3D tomography reveals connections between the phagophore and endoplasmic reticulum. Autophagy (2009) 5:1180-5.

[27] Simonsen A & Tooze SA. Coordination of membrane events during autophagy by multiple class III PI3-kinase complexes. J Cell Biol (2009) 186:773-82.

[28] Tooze SA, Yoshimori T. The origin of the autophagosomal membrane. Nat Cell Biol (2010) 12:831-5.

[29] Hailey DW, Rambold AS, Satpute-Krishnan P, Mitra K, Sougrat R, Kim PK, Lippincott-Schwartz J. Mitochondria supply membranes for autophagosome biogenesis during starvation. Cell (2010) 141:656-67.

[30] Van der Vaart A, Reggiori F. The Golgi complex as a source for yeast autophagosomal membranes. Autophagy (2010) 6:800-1.

[31] Yen WL, Shintani T, Nair U, Cao Y, Richardson BC, Li Z, Hughson FM, Baba M, Klionsky DJ. The conserved oligomeric Golgi complex is involved in double-membrane vesicle formation during autophagy. J Cell Biol (2010) 188:101-14.

[32] Ravikumar B, Moreau K, Jahreiss L, Puri C, Rubinsztein DC. Plasma membrane contributes to the formation of pre-autophagosomal structures. Nat Cell Biol 12:747-57. Erratum in: Nat Cell Biol (2010) 12:1021.

[33] Novikoff PM, Novikoff AB, Quintana N, Hauw JJ. Golgi apparatus, GERL, and lysosomes of neurons in rat dorsal root ganglia, studied by thick section and thin section cytochemistry. J Cell Biol (1971) 50:859-86.

[34] Broadwell RD, Cataldo AM. The neuronal endoplasmic reticulum: its cytochemistry and contribution to the endomembrane system. II. Axons and terminals. J Comp Neurol (1984) 230:231-48.

[35] Matthews MR, Raisman G. A light and electron microscopic study of the cellular response to axonal injury in the superior cervical ganglion of the rat. Proc R Soc Lond B Biol Sci (1972) 181:43-79.

[36] Wang QJ, Ding Y, Kohtz DS, Mizushima N, Cristea IM, Rout MP, Chait BT, Zhong Y, Heintz N, Yue Z. Induction of autophagy in axonal dystrophy and degeneration. J Neurosci (2006) 26:8057-68.

[37] Yue Z. Regulation of neuronal autophagy in axon. Implication of autophagy in axonal function and dysfunction/degeneration. Autophagy (2007) 3:139-141.

[38] Liang XH, Jackson S, Seaman M, Brown K, Kempkes B, Hibshoosh H, Levine B. Induction of autophagy and inhibition of tumorigenesis by beclin-1. Nature (1999) 402:672-6.

[39] Selimi F, Lohof AM, Heitz S, Lalouette A, Jarvis CI, Bailly Y, Mariani J. Lurcher GRID2-induced death and depolarization can be dissociated in cerebellar Purkinje cells. Neuron (2003) 37:813-9.

[40] Yue Z, Horton A, Bravin M, DeJager PL, Selimi F, Heintz N. A novel protein complex linking the delta 2 glutamate receptor and autophagy: implications for neurodegeneration in lurcher mice. Neuron (2002) 35:921-33.

[41] Yang DS, Stavrides P, Mohan PS, Kaushik S, Kumar A, Ohno M, Schmidt SD, Wesson DW, Bandyopadhyay U, Jiang Y, Pawlik M, Peterhoff CM, Yang AJ, Wilson DA, St George-Hyslop P, Westaway D, Mathews PM, Levy E, Cuervo AM, Nixon RA. Therapeutic effects of remediating autophagy failure in a mouse model of Alzheimer disease by enhancing lysosomal proteolysis. Autophagy (2011) 7:788-9.

[42] Jarheiss L, Menzies FM, Rubinsztein DC. The itinerary of autophagosomes: from peripheral formation to kiss-and-run fusion with lysosomes. Traffic (2008) 9:574-587.

[43] Halpain J, Dehmelt L. The MAP1 family of microtubule-associated proteins Genome biology (2006) 7:224-230.

[44] Köchl R, Hu XW, Chan EY, Tooze SA. Microtubules facilitate autophagosome formation and fusion of autophagosomes with endosomes. Traffic (2006) 7:129-45.

[45] Fass E, Shvets E, Degani I, Hirschberg K, Elazar Z. Microtubules support production of starvation-induced autophagosomes but not their targeting and fusion with lysosomes. J Biol Chem (2006) 281:36303-16.

[46] Gonzalez-Billault C, Jimenez-Mateos EM, Caceres A, Diaz-Nido J, Wandosell F, Avila J. Microtubule-associated protein 1B function during normal development, regeneration, and pathological conditions in the nervous system. J Neurobiol (2004) 58:48-59.

[47] Nixon RA, Wegiel J, Kumar A, Yu WH, Peterhoff C, Cataldo A, Cuervo AM. Extensive involvement of autophagy in Alzheimer disease: an immuno-electron microscopy study. J Neuropathol Exp Neurol (2005) 64:113-22.

[48] Boland B & Nixon RA. Neuronal macroautophagy: from development to degeneration. Mol Aspects Med (2006) 27:503-19.

[49] Boland B, Kumar A, Lee S, Platt FM, Wegiel J, Yu WH, Nixon RA. Autophagy induction and autophagosome clearance in neurons: relationship to autophagic pathology in Alzheimer's disease. J Neurosci (2008) 28:6926-37.

[50] Lee S, Sato Y, Nixon RA. Primary lysosomal dysfunction causes cargo-specific deficits of axonal transport leading to Alzheimer-like neuritic dystrophy. Autophagy (2011a) 7:1562-1563.

[51] Sigmond T, Fehér J, Baksa A, Pásti G, Pálfia Z, Takács-Vellai K, Kovács J, Vellai T, Kovács AL. Qualitative and quantitative characterization of autophagy in Caenorhabditis elegans by electron microscopy. Methods Enzymol (2008) 451:467-91.

[52] Okazaki N, Yan J, Yuasa S, Ueno T, Kominami E, Masuho Y, Koga H, Muramatsu M. Interaction of the Unc-51-like kinase and microtubule-associated protein light chain 3 related proteins in the brain: possible role of vesicular transport in axonal elongation. Brain Res Mol Brain Res (2000) 85:1-12.

[53] Komatsu M, Wang QJ, Holstein GR, Friedrich VL Jr, Iwata J, Kominami E, Chait BT, Tanaka K, Yue Z. Essential role for autophagy protein Atg7 in the maintenance of axonal homeostasis and the prevention of axonal degeneration. Proc Natl Acad Sci USA (2007a) 104:14489-94.

[54] Nishiyama J, Miura E, Mizushima N, Watanabe M, Yuzaki M. Aberrant membranes and double-membrane structures accumulate in the axons of Atg5-null Purkinje cells before neuronal death. Autophagy (2007) 3:591-6.

[55] Komatsu M, Waguri S, Koike M, Sou YS, Ueno T, Hara T, Mizushima N, Iwata J, Ezaki J, Murata S, Hamazaki J, Nishito Y, Iemura S, Natsume T, Yanagawa T, Uwayama J, Warabi E, Yoshida H, Ishii T, Kobayashi A, Yamamoto M, Yue Z, Uchiyama Y, Kominami E, Tanaka K. Homeostatic levels of p62 control cytoplasmic inclusion body formation in autophagy-deficient mice. Cell (2007b) 131:1149-63.

[56] Pankiv S, Clausen TH, Lamark T, Brech A, Bruun JA, Outzen H, Øvervatn A, Bjørkøy G, Johansen T. p62/SQSTM1 binds directly to Atg8/LC3 to facilitate degradation of ubiquitinated protein aggregates by autophagy. J Biol Chem (2007) 282:24131-45.

[57] Zhou X, Babu JR, da Silva S, Shu Q, Graef IA, Oliver T, Tomoda T, Tani T, Wooten M, Wang F. Unc-51-like kinase 1/2-mediated endocytic processes regulate filopodia extension and branching of sensory axons. Proc Natl Acad Sci USA (2007) 104:5842-7.

[58] Coleman M. Axon degeneration mechanisms: commonality amid diversity. Nat Rev Neurosci (2005) 6:889-98.

[59] Raff MC, Whitmore AV, Finn JT. Axonal self-destruction and neurodegeneration. Science (2002) 296:868-71.

[60] Yang Y, Fukui K, Koike T, Zheng X. Induction of autophagy in neurite degeneration of mouse superior cervical ganglion neurons. Eur J Neurosci (2007) 26:2979-2988.

[61] Shehata M, Matsumura H, Okubo-Suzuki R, Ohkawa N, Inokuchi K. Neuronal stimulation induces autophagy in hippocampal neurons that is involved in AMPA receptor degradation after chemical long-term depression. J Neurosci (2012) 32:10413-22.

[62] Gordon PB, Seglen PO. Prelysosomal convergence of autophagic and endocytic pathways. Biochem Biophys Res Commun (1988) 151:40-7.

[63] Rusten TE, Vaccari T, Lindmo K, Rodahl LM, Nezis IP, Sem-Jacobsen C, Wendler F, Vincent JP, Brech A, Bilder D, Stenmark H. ESCRTs and Fab1 regulate distinct steps of autophagy. Curr Biol (2007) 17:1817-25.

[64] Filimonenko M, Stuffers S, Raiborg C, Yamamoto A, Malerød L, Fisher EM, Isaacs A, Brech A, Stenmark H, Simonsen A. Functional multivesicular bodies are required for autophagic clearance of protein aggregates associated with neurodegenerative disease. J Cell Biol (2007) 179:485-500.

[65] Larsen KE, Fon EA, Hastings TG, Edwards RH, Sulzer D. Methamphetamine-induced degeneration of dopaminergic neurons involves autophagy and upregulation of dopamine synthesis. J Neurosci (2002) 22:8951-60.

[66] Ravikumar B, Duden R, Rubinsztein DC. Aggregate-prone proteins with polyglutamine and polyalanine expansions are degraded by autophagy. Hum Mol Genet (2002) 11:1107-17.

[67] Rubinsztein DC. The roles of intracellular protein-degradation pathways in neurodegeneration. Nature (2006) 443:780-6.

[68] Rubinsztein DC, Gestwicki JE, Murphy LO, Klionsky DJ. Potential therapeutic applications of autophagy. Nat Rev Drug Discov (2007) 6:304-12.

[69] Nezis IP, Simonsen A, Sagona AP, Finley K, Gaumer S, Contamine D, Rusten TE, Stenmark H, Brech A. Ref(2)P, the Drosophila melanogaster homologue of mammalian p62, is required for the formation of protein aggregates in adult brain. J Cell Biol (2008) 180:1065-71.

[70] Collins CA, Wairkar YP, Johnson SL, DiAntonio A. Highwire restrains synaptic growth by attenuating a MAP kinase signal. Neuron (2006) 51:57-69.

[71] Shen W, Ganetzky B. Autophagy promotes synapse development in Drosophila. J Cell Biol (2009) 187:71-9.

[72] Wan HI, DiAntonio A, Fetter RD, Bergstrom K, Strauss R, Goodman CS. Highwire regulates synaptic growth in Drosophila. Neuron (2000) 26:313-29.

[73] Rowland AM, Richmond JE, Olsen JG, Hall DH, Bamber BA. Presynaptic terminals independently regulate synaptic clustering and autophagy of GABAA receptors in Caenorhabditis elegans. J Neurosci (2006) 26:1711-20.

[74] Zhang XD, Wang Y, Wang Y, Zhang X, Han R, Wu JC, Liang ZQ, Gu ZL, Han F, Fukunaga K, Qin ZH. p53 mediates mitochondria dysfunction-triggered autophagy activation and cell death in rat striatum. Autophagy (2009) 5:339-50.

[75] Wairkar YP, Toda H, Mochizuki H, Furukubo-Tokunaga K, Tomoda T, Diantonio A. Unc-51 controls active zone density and protein composition by downregulating ERK signaling. J Neurosci (2009) 29:517-28.

[76] Hernandez D, Torres CA, Setlik W, Cebrián C, Mosharov EV, Tang G, Cheng HC, Kholodilov N, Yarygina O, Burke RE, Gershon M, Sulzer D. Regulation of presynaptic neurotransmission by macroautophagy. Neuron (2012) 74:277–284.

[77] Kamada Y, Yoshino K, Kondo C, Kawamata T, Oshiro N, Yonezawa K, Ohsumi Y. Tor directly controls the Atg1 kinase complex to regulate autophagy. Mol Cell Biol (2010) 30:1049-58.

[78] Huang J & Manning BD. A complex interplay between Akt, TSC2 and the two mTOR complexes. Biochem Soc Trans (2009) 37:217-22.

[79] Long X, Müller F, Avruch J. TOR action in mammalian cells and in Caenorhabditis elegans. Curr Top Microbiol Immunol (2004) 279:115-38.

[80] Richter JD, Klann E. Making synaptic plasticity and memory last: mechanisms of translational regulation. Genes Dev (2009) 23:1-11.

[81] Hu JY, Chen Y, Schacher S. Protein kinase C regulates local synthesis and secretion of a neuropeptide required for activity-dependent long-term synaptic plasticity. J Neurosci (2007) 27:8927-8939.

[82] Weragoda RM, Walters ET. Serotonin induces memory-like, rapamycin-sensitive hyperexcitability in sensory axons of aplysia that contributes to injury responses. J Neurophysiol (2007) 98:1231-9.

[83] Lee S, Sato Y, Nixon RA. Lysosomal proteolysis inhibition selectively disrupts axonal transport of degradative organelles and causes an Alzheimer's-like axonal dystrophy. J Neurosci (2011) 31(21):7817-30.

[84] Bunge MB. Fine structure of nerve fibers and growth cones of isolated sympathetic neurons in culture. J Cell Biol (1973) 56:713-35.

[85] Zanjani SH, Selimi F, Vogel MW, Haeberlé AM, Boeuf J, Mariani J, Bailly YJ. Survival of interneurons and parallel fiber synapses in a cerebellar cortex deprived of Purkinje cells: studies in the double mutant mouse Grid2Lc/+;Bax(-/-). J Comp Neurol (2006) 497:622-35.

[86] Eskelinen EL. Maturation of autophagic vacuoles in mammalian cells. Autophagy (2005) 1:1-10.

[87] Mizushima N, Klionsky DJ.Protein turnover via autophagy: implications for metabolism. Annu Rev Nutr (2007) 27:19-40.

[88] Ramesh Babu J, Lamar Seibenhener M, Peng J, Strom AL, Kemppainen R, Cox N, Zhu H, Wooten MC, Diaz-Meco MT, Moscat J, Wooten MW. Genetic inactivation of p62 leads to accumulation of hyperphosphorylated tau and neurodegeneration. J Neurochem (2008) 106:107-20.

[89] Jiang J, Parameshwaran K, Seibenhener ML, Kang MG, Suppiramaniam V, Huganir RL, Diaz-Meco MT, Wooten MW. AMPA receptor trafficking and synaptic plasticity require SQSTM1/p62. Hippocampus (2009) 19:392-406.

[90] Jobim PF, Pedroso TR, Werenicz A, Christoff RR, Maurmann N, Reolon GK, Schröder N, Roesler R. Impairment of object recognition memory by rapamycin inhibition of mTOR in the amygdala or hippocampus around the time of learning or reactivation. Behav Brain Res (2012) 228:151-8.

[91] Cota D, Proulx K, Smith KA, Kozma SC, Thomas G, Woods SC, Seeley RJ. Hypothalamic mTOR signaling regulates food intake. Science (2006) 312:927-30.

[92] Young JE, La Spada AR. Development of selective nutrient deprivation as a system to study autophagy induction and regulation in neurons. Autophagy (2009) 5:555-7.

[93] Sarkar S, Ravikumar B, Floto RA, Rubinsztein DC. Rapamycin and mTOR-independent autophagy inducers ameliorate toxicity of polyglutamine-expanded huntingtin and related proteinopathies. Cell Death Differ (2009) 16(1):46-56.

[94] Du L, Hickey RW, Bayir H, Watkins SC, Tyurin VA, Guo F, Kochanek PM, Jenkins LW, Ren J, Gibson G, Chu CT, Kagan VE, Clark RS. Starving neurons show sex difference in autophagy. J Biol Chem (2009) 284:2383-96.

[95] Koike M, Shibata M, Tadakoshi M, Gotoh K, Komatsu M, Waguri S, Kawahara N, Kuida K, Nagata S, Kominami E, Tanaka K, Uchiyama Y. Inhibition of autophagy prevents hippocampal pyramidal neuron death after hypoxic-ischemic injury. Am J Pathol (2008) 172:454-69.

[96] Adhami F, Liao G, Morozov YM, Schloemer A, Schmithorst VJ, Lorenz JN, Dunn RS, Vorhees CV, Wills-Karp M, Degen JL, Davis RJ, Mizushima N, Rakic P, Dardzinski BJ, Holland SK, Sharp FR, Kuan CY. Cerebral ischemia-hypoxia induces intravascular coagulation and autophagy. Am J Pathol (2006) 169:566-83.

[97] Borsello T, Croquelois K, Hornung JP, Clarke PG. N-methyl-d-aspartate-triggered neuronal death in organotypic hippocampal cultures is endocytic, autophagic and mediated by the c-Jun N-terminal kinase pathway. Eur J Neurosci (2003) 18:473-485.

[98] Wang Y, han R, Liang ZQ, Wu JC,ZhangXD, Gu ZI, Qin ZH. An autophagic mechanism is involved in apoptotic death of rat striatal neurons induced by the non N-methyl-D-aspartate receptor agonist kainic acid. Autophagy (2008) 4:214-226.

[99] Høyer-Hansen M, Jäättelä M. Connecting endoplasmic reticulum stress to autophagy by unfolded protein response and calcium. Cell Death Differ (2007) 14:1576-82.

[100] Castino R, Lazzeri G, Lenzi P, Bellio N, Follo C, Ferrucci M, Fornai F, Isidoro C. Suppression of autophagy precipitates neuronal cell death following low doses of methamphetamine. J Neurochem (2008) 106:1426-39.

[101] Zhu JH, Horbinsky C, Guo F, Watkins S, Uchiyama Y, Chu CT. Regulation of autophagy by extracellular signal-regulated protein kinases during 1-methyl-4-phenylpyridinium-induced cell death. Am J Pathol (2007) 170:75-86.

[102] Ding Q, Dimayuga E, Martin S, Bruce-Keller AJ, Nukala V, Cuervo AM, Keller JN. Characterization of chronic low-level proteasome inhibition on neural homeostasis. J Neurochem (2003) 86:489-97.

[103] Pandey UB, Nie Z, Batlevi Y, McCray BA, Ritson GP, Nedelsky NB, Schwartz SL, DiProspero NA, Knight MA, Schuldiner O, Padmanabhan R, Hild M, Berry DL, Garza D, Hubbert CC, Yao TP, Baehrecke EH, Taylor JP. HDAC6 rescues neurodegeneration and provides an essential link between autophagy and the UPS. Nature (2007) 447:859-63.

[104] Bedford L, Hay D, Devoy A, Paine S, Powe DG, Seth R, Gray T, Topham I, Fone K, Rezvani N, Mee M, Soane T, Layfield R, Sheppard PW, Ebendal T, Usoskin D, Lowe J, Mayer RJ. Depletion of 26S proteasomes in mouse brain neurons causes neurodegeneration and Lewy-like inclusions resembling human pale bodies. J Neurosci (2008) 28:8189-98.

[105] Koike M, Shibata M, Waguri S, Yoshimura K, Tanida I, Kominami E, Gotow T, Peters C, von Figura K, Mizushima N, Saftig P, Uchiyama Y. Participation of autophagy in storage of lysosomes in neurons from mouse models of neuronal ceroid-lipofuscinoses (Batten disease). Am J Pathol (2005) 167:1713-28.

[106] Liao G, Yao Y, Liu J, Yu Z, Cheung S, Xie A, Liang X, Bi X. Cholesterol accumulation is associated with lysosomal dysfunction and autophagic stress in Npc1 -/- mouse brain. Am J Pathol (2007) 171:962-75.

[107] Pacheco CD, Lieberman AP. Lipid trafficking defects increase Beclin-1 and activate autophagy in Niemann-Pick type C disease. Autophagy (2007) 3:487-9.

[108] Vergarajauregui S, Connelly PS, Daniels MP, Puertollano R. Autophagic dysfunction in mucolipidosis type IV patients. Hum Mol Genet (2008) 17:2723-2737.

[109] Clausen TH, Lamark T, Isakson P, Finley K, Larsen KB, Brech A, Øvervatn A, Stenmark H, Bjørkøy G, Simonsen A, Johansen T. p62/SQSTM1 and ALFY interact to facilitate the formation of p62 bodies/ALIS and their degradation by autophagy. Autophagy (2010) 6:330-44.

[110] Knaevelsrud H, Simonsen A. Fighting disease by selective autophagy of aggregate-prone proteins. FEBS Lett (2010) 584:2635-45.

[111] Webb JL, Ravikumar B, Atkins J, Skepper JN, Rubinsztein DC. Alpha-synuclein is degraded by both autophagy and the proteasome. J Biol Chem (2003) 278:25009-25013.

[112] Fortun J, Dunn WA Jr, Joy S, Li J, Notterpek L. Emerging role for autophagy in the removal of aggresomes in Schwann cells. J Neurosci (2003) 23:10672-80.

[113] Bjørkøy G, Lamark T, Brech A, Outzen H, Perander M, Overvatn A, Stenmark H, Johansen T. p62/SQSTM1 forms protein aggregates degraded by autophagy and has a protective effect on huntingtin-induced cell death. J Cell Biol (2005) 171:603-14.

[114] Kabuta T, Suzuki Y, Wada K. Degradation of amyotrophic lateral sclerosis-linked mutant Cu/Zn-superoxide dismutase proteins by macroautophagy and the proteasome. J Biol Chem (2006) 281:30524-33.

[115] Berger Z, Ravikumar B, Menzies FM, Oroz LG, Underwood BR, Pangalos MN, Schmitt I, Wullner U, Evert BO, O'Kane CJ, Rubinsztein DC. Rapamycin alleviates toxicity of different aggregate-prone proteins. Hum Mol Genet (2006) 15:433-42.

[116] Petersén A, Larsen KE, Behr GG, Romero N, Przedborski S, Brundin P, Sulzer D. Expanded CAG repeats in exon 1 of the Huntington's disease gene stimulate dopamine-mediated striatal neuron autophagy and degeneration. Hum Mol Genet (2001) 10:1243-54.

[117] Rubinsztein DC, DiFiglia M, Heintz N, Nixon RA, Qin ZH, Ravikumar B, Stefanis L, Tolkovsky A. Autophagy and its possible roles in nervous system diseases, damage and repair. Autophagy (2005) 1:11-22.

[118] Ventruti A, Cuervo AM. Autophagy and neurodegeneration. Curr Neurol Neurosci Rep (2007) 7:443-51.

[119] Nixon RA, Yang DS, Lee JH. Neurodegenerative lysosomal disorders: a continuum from development to late age. Autophagy (2008) 4:590-9.

[120] Lee JH, Yu WH, Kumar A, Lee S, Mohan PS, Peterhoff CM, Wolfe DM, Martinez-Vicente M, Massey AC, Sovak G, Uchiyama Y, Westaway D, Cuervo AM, Nixon RA. Lysosomal proteolysis and autophagy require presenilin 1 and are disrupted by Alzheimer-related PS1 mutations. Cell (2010) 141:1146-1158.

[121] Mariño G, Madeo F, Kroemer G. Autophagy for tissue homeostasis and neuroprotection. Curr Opin Cell Biol (2011) 23:198-206.

[122] Bence NF, Sampat RM, Kopito RR. Impairment of the ubiquitin-proteasome system by protein aggregation. Science (2001) 292:1552-5.

[123] Fleming A, Noda T, Yoshimori T, Rubinsztein DC. Chemical modulators of autophagy as biological probes and potential therapeutics. Nat Chem Biol (2011) 7:9-17.

[124] Nixon RA. Autophagy, amyloidogenesis and Alzheimer disease. J Cell Sci (2007) 120:4081-91.

[125] Spilman P, Podlutskaya N, Hart MJ, Debnath J, Gorostiza O, Bredesen D, Richardson A, Strong R, Galvan V. Inhibition of mTOR by rapamycin abolishes cognitive deficits

and reduces amyloid-beta levels in a mouse model of Alzheimer's disease. PLoS One (2010) 5:e9979.

[126] Heiseke A, Aguib Y, Riemer C, Baier M, Schatzl HM. Lithium induces clearance of protease resistant prion protein in prion-infected cells by induction of autophagy. J Neurochem (2009) 109:25-34.

[127] Cortes CJ, Qin K, Cook J, Solanki A, Mastrianni JA. Rapamycin delays disease onset and prevents PrP plaque deposition in a mouse model of Gerstmann-Sträussler-Scheinker disease. J Neurosci (2012) 32:12396-12405.

[128] Massey AC, Kaushik S, Cuervo AM. Lysosomal chat maintains the balance. Autophagy (2006) 2:325-7.

[129] Iwata A, Riley BE, Johnston JA, Kopito RR. HDAC6 and microtubules are recquired for autophagic degradation of aggregated huntingtin. J Biol Chem (2005) 280:40282-40292.

[130] Martinez-Vicente M, Cuervo AM. Autophagy and neurodegeneration: when the cleaning crew goes on strike. Lancet Neurol (2007) 6:352-61.

[131] Shimizu S, Kanaseki T, Mizushima N, Mizuta T, Arakawa-Kobayashi S, Thompson CB, Tsujimoto Y. Role of Bcl-2 family proteins in a non-apoptotic programmed cell death dependent on autophagy genes. Nat Cell Biol (2004) 6:1221-8.

[132] Veneault-Fourrey C, Talbot NJ. Autophagic cell death and its importance for fungal developmental biology and pathogenesis. Autophagy(2007) 3:126-7.

[133] Yu WH, Kumar A, Peterhoff C, Shapiro Kulnane L, Uchiyama Y, Lamb BT, Cuervo AM, Nixon RA. Autophagic vacuoles are enriched in amyloid precursor protein-secretase activities: implications for beta-amyloid peptide over-production and localization in Alzheimer's disease. Int J Biochem Cell Biol (2004) 36:2531-40.

[134] Yu L, Wan F, Dutta S, Welsh S, Liu Z, Freundt E, Baehrecke EH, Lenardo M. Autophagic programmed cell death by selective catalase degradation. Proc Natl Acad Sci USA (2006) 103:4952-7.

[135] Scott RC, Juhász G, Neufeld TP. Direct induction of autophagy by Atg1 inhibits cell growth and induces apoptotic cell death. Curr Biol (2007) 17:1-11.

[136] Wullschleger S, Loewith R, Hall MN. TOR signaling in growth and metabolism. Cell (2006) 124:471-84.

[137] Arico S, Petiot A, Bauvy C, Dubbelhuis PF, Meijer AJ, Codogno P, Ogier-Denis E. The tumor suppressor PTEN positively regulates macroautophagy by inhibiting the phosphatidylinositol 3-kinase/protein kinase B pathway. J Biol Chem (2001) 276:35243-6.

[138] Jin S. Autophagy, mitochondrial quality control, and oncogenesis. Autophagy (2006) 2:80-4.

[139] Martin DN, Baehrecke EH. Caspases function in autophagic programmed cell death in Drosophila. Development (2004) 131:275-84.

[140] Chu CT. Autophagic stress in neuronal injury and disease. J Neuropathol Exp Neurol (2006) 65:423-32.

[141] Alirezaei M, Jelodar G, Niknam P, Ghayemi Z, Nazifi S. Betaine prevents ethanol-induced oxidative stress and reduces total homocysteine in the rat cerebellum. J Physiol Biochem (2011) 67:605-12.

[142] Prusiner SB. Prions. Proc Natl Acad Sci USA(1998) 95:13363-13383.

[143] Weissmann C. The state of prion. Nat Rev Microbiol (2004) 2:861-871.

[144] Aguzzi A, Polymenidou M. Mammalian prion biology: one century of evolving concepts. Cell (2004) 116:313-327.

[145] Collinge J. Molecular neurology of prion disease. J Neurol Neurosurg Psychiatry (2005) 76:906-919.

[146] Prusiner SB. Novel proteinaceous infectious particles cause scrapie. Science (1982) 216:136-144.

[147] Aguzzi A, Haass C.Games played by rogue proteins in prion disorders and Alzheimer's disease. Science (2003) 302:814-818.

[148] Brandner S, Raeber A, Sailer A, Blättler T, Fischer M, Weissmann C, Aguzzi A. Normal host prion protein (PrPC) is required for scrapie spread within the central nervous system. Proc Natl Acad Sci USA (1996) 93:13148-51.

[149] Büeler H, Aguzzi A, Sailer A, Greiner RA, Autenried P, Aguet M, Weissmann C. Mice devoid of PrP are resistant to scrapie. Cell (1993) 73:1339-47.

[150] Radford HE, Mallucci GR. The role of GPI-anchored PrP(C) in mediating the neurotoxic effect of scrapie prions in neurons. Curr Issues Mol Biol (2010) 12:119-128.

[151] Harris DA, True HL. New insights into prion structure and toxicity. Neuron (2006) 50:353-7.

[152] Winklhofer KF, Tatzelt J, Haass C. The two faces of protein misfolding: gain- and loss-of-function in neurodegenerative diseases. EMBO J (2008) 27:336-49.

[153] Brown DR, Qin K, Herms JW, Madlung A, Manson J, Strome R, Fraser PE, Kruck T, von Bohlen A, Schulz-Schaeffer W, Giese A, Westaway D, Kretzschmar H. The cellular prion protein binds copper in vivo. Nature (1997) 390:684-7.

[154] Gauczynski S, Peyrin JM, Haïk S, Leucht C, Hundt C, Rieger R, Krasemann S, Deslys JP, Dormont D, Lasmézas CI, Weiss S. The 37-kDa/67-kDa laminin receptor acts as the cell-surface receptor for the cellular prion protein. Embo J (2001) 20:5863-75.

[155] Schmitt-Ulms G, Legname G, Baldwin MA, Ball HL, Bradon N, Bosque PJ, Crossin KL, Edelman GM, DeArmond SJ, Cohen FE, Prusiner SB. Binding of neural cell adhesion molecules (N-CAMs) to the cellular prion protein. J Mol Biol (2001) 314:1209-25.

[156] Loubet D, Dakowski C, Pietri M, Pradines E, Bernard S, Callebert J, Ardila-Osorio H, Mouillet-Richard S, Launay JM, Kellermann O, Schneider B. Neuritogenesis: the prion protein controls b1 integrin signaling activity. FASEB J (2012) 26:678-90.

[157] Brown DR, Schulz-Schaeffer WJ, Schmidt B, Kretzschmar HA. Prion protein-deficient cells show altered response to oxidative stress due to decreased SOD-1 activity. Exp Neurol (1997) 146:104-12.

[158] Kuwahara C, Takeuchi AM, Nishimura T, Haraguchi K, Kubosaki A, Matsumoto Y, Saeki K, Matsumoto Y, Yokoyama T, Itohara S, Onodera T.Prions prevent neuronal cell-line death. Nature (1999) 400:225-6.

[159] Milhavet O, Lehmann S. Oxidative stress and the prion protein in transmissible spongiform encephalopathies. Brain Res Brain Res Rev (2002) 38:328-39.

[160] Spudich A, Frigg R, Kilic E, Kilic U, Oesch B, Raeber A, Bassetti CL, Hermann DM. Aggravation of ischemic brain injury by prion protein deficiency: role of ERK-1/-2 and STAT-1. Neurobiol Dis (2005) 20:442-449.

[161] Weise J, Sandau R, Schwarting S, Crome O, Wrede A, Schulz-Schaeffer W, Zerr I, Bähr M. Deletion of cellular prion protein results in reduced Akt activation, enhanced postischemic caspase-3 activation, and exacerbation of ischemic brain injury. Stroke (2006) 37:1296-300.

[162] Mitteregger G, Vosko M, Krebs B, Xiang W, Kohlmannsperger V, Nölting S, Hamann GF, Kretzschmar HA. The role of the octarepeat region in neuroprotective function of the cellular prion protein. Brain Pathol (2007) 17:174-83.

[163] Zanata SM, Lopes MH, Mercadante AF, Hajj GN, Chiarini LB, Nomizo R, Freitas AR, Cabral AL, Lee KS, Juliano MA, de Oliveira E, Jachieri SG, Burlingame A, Huang L, Linden R, Brentani RR, Martins VR. Stress-inducible protein 1 is a cell surface ligand for cellular prion that triggers neuroprotection. Embo J (2002) 21:3307-16.

[164] Schneider B, Mutel V, Pietri M, Ermonval M, Mouillet-Richard S, Kellermann O. NADPH oxidase and extracellular regulated kinases 1/2 are targets of prion protein signaling in neuronal and nonneuronal cells. Proc Natl Acad Sci USA (2003) 100:13326-31.

[165] Pradines E, Loubet D, Schneider B, Launay JM, Kellermann O, Mouillet-Richard S. CREB-dependent gene regulation by prion protein: impact on MMP-9 and beta-dystroglycan. Cell Signal (2008) 20:2050-2058.

[166] Pradines E, Loubet D, Mouillet-Richard S, Manivet P, Launay JM, Kellermann O, Schneider B. Cellular prion protein coupling to TACE-dependent TNF-alpha shed-

ding controls neurotransmitter catabolism in neuronal cells. J Neurochem (2009) 110:912-23.

[167] Linden R, Martins VR, Prado MA, Cammarota M, Izquierdo I, Brentani RR. Physiology of the prion protein. Physiol Rev (2008) 88:673-728.

[168] Moreno JA, Radford H, Peretti D, Steinert JR, Verity N, Martin MG, Halliday M, Morgan J, Dinsdale D, Ortori CA, Barrett DA, Tsaytler P, Bertolotti A, Willis AE, Bushell M, Mallucci GR. Sustained translational repression by eIF2α-P mediates prion neurodegeneration. Nature (2012) 485:507-511.

[169] Ashe KH, Aguzzi A. Prions, prionoids and pathogenic proteins in Alzheimer disease. Prion (2013) 7: in press.

[170] Kristiansen M, Deriziotis P, Dimcheff DE, Jackson GS, Ovaa H, Naumann H, Clarke AR, van Leeuwen FW, Menéndez-Benito V, Dantuma NP, Portis JL, Collinge J, Tabrizi SJ. Disease-associated prion protein oligomers inhibit the 26S proteasome. Mol Cell (2007) 26:175-88.

[171] Pietri M, Caprini A, Mouillet-Richard S, Pradines E, Ermonval M, Grassi J, Kellermann O, Schneider Schneider B. Overstimulation of PrPC signaling pathways by prion peptide 106-126 causes oxidative injury of bioaminergic neuronal cells. J Biol Chem (2006) 281:28470-9.

[172] Heiseke A, Aguib Y, Schatzl HM. Autophagy, prion infection and their mutual interactions. Curr Issues Mol Biol (2010) 12:87-98.

[173] Sikorska B, Liberski PP, Giraud P, Kopp N, Brown P. Autophagy is a part of ultrastructural synaptic pathology in Creutzfeldt-Jakob disease: a brain biopsy study. Int J Biochem Cell Biol (2004) 36:2563-73.

[174] Liberski PP, Brown DR, Sikorska B, Caughey B, Brown P. Cell death and autophagy in prion diseases (transmissible spongiform encephalopathies). Folia Neuropathol (2008) 46:1-25.

[175] Boellaard JW, Schlote W, Tateishi J. Neuronal autophagy in experimental Creutzfeldt-Jakob's disease. Acta Neuropathol (1989) 78:410-418.

[176] Boellaard JW, Kao M, Schlote W, Diringer H. Neuronal autophagy in experimental scrapie. Acta Neuropathol (1991) 82:225-228.

[177] Schätzl HM, Laszlo L, Holtzman DM, Tadzelt J, DeArmond SJ, Weiner RI, Mobley WC, Prusiner SB. A hypothalamic neuronal cell line persistently infected with scrapie prions exhibits apoptosis. J Virol (1997) 71:8821-8831.

[178] Dron M, Dandoy-Dron F, Guillo F, Benboudjema L, Haw J-J, Lebon P, Dormont D, Tovey MG. Characterization of the human analogue of a scrapie-responsive gene. J Biol Chem (1998) 273:18015-18018.

[179] Dandoy-Dron F, Guillo F, Benboudjema L, Deslys J-P, Lasmézas C, Dormont D, Tovey MG, Dron M. Gene expression in scrapie. Cloning of a new scrapie-responsive

gene and the identification of seven other mRNA transcripts. J Biol Chem (1998) 273:7691-7697.

[180] Dandoy-Dron F, Benboudjema L, Guillo F, Jaegly A, Jasmin C, Dormont D, Tovey MG, Dron M. Enhanced levels of scrapie responsive gene mRNA in BSE-infected mouse brain. Brain Res Mol Brain Res (2000) 76:173-179.

[181] Dron M, Bailly Y, Beringue V, Haeberlé A-M, Griffond B, Risold P-Y, Tovet MG, Laude H, Dandoy-Dron F. Scrg1 is induced in TSE and brain injuries, and associated with autophagy. Eur J Neurosci (2005) 22:133-146.

[182] Dron M, Bailly Y, Beringue V, Haeberlé A-M, Griffond B, Risold P-Y, Tovey MG, Laude H, Dandoy-Dron F. SCRG1, a potential marker of autophagy in transmisible spongiform encephalopathies. Autophagy (2006) 2:58-60.

[183] Xu Y, Tian C, Wang SB, Xie WL, Guo Y, Zhang J, Shi Q, Chen C, Dong XP. Activation of the macroautophagic system in scrapie-infected experimental animals and human genetic prion diseases. Autophagy (2012) 8 (in press).

[184] Hung SY, Huang WP, Liou HC, Fu WM. Autophagy protects neurons from Abeta-induced cytotoxicity. Autophagy (2009) 5:502-510.

[185] Doh-Ura K, Iwaki T, Caughey B. Lysosomotropic agents and cysteine protease inhibitors inhibit scrapie-associated prion protein accumulation. J Virol (2000) 74:4894-4897.

[186] Ertmer A, Gilch S, Yun SW, Flechsig E, Klebl B, Stein-Gerlach M, Klein MA, Schätzl HM. The tyrosine kinase inhibitor STI571 induces cellular clearance of PrPsc in prion-infected cells. J Biol Chem (2004) 279:41918-41927.

[187] Oh JM, Shin HY, Park SJ, Kim BH, Choi JK, Choi EK, Carp RI, Kim YS. The involvement of cellular prion protein in the autophagy pathway in neuronal cells. Mol Cell Neurosci (2008) 39:238-47.

[188] Mok SW, Riemer C, Madela K, Hsu DK, Liu FT, Gültner S, Heise I, Baier M. Role of galectin-3 in prion infections of the CNS. Biochem Biophys Res Commun (2007) 359:672-8.

[189] Ertmer A, Huber V, Gilch S, Yoshimori T, Erfle V, Duyster J, Elsässer HP, Schätzl HM. The anticancer drug imatinib induces cellular autophagy. Leukemia (2007) 21:936-942.

[190] Yun SW, Ertmer A, Flechsig E, Gilch S, Riederer P, Gerlach M, Schatzl HM, Klein MA. The tyrosine kinase inhibitor imatinib mesylate delays prion neuroinvasion by inhibiting prion propagation in the periphery. J Neurovirol (2007) 13:328-337.

[191] Sarkar S, Floto RA, Berger Z, Imarisio S, Cordenier A, Pasco M, Cook LJ, Rubinsztein DC. Lithium induces autophagy by inhibiting inositol monophosphatase. J Cell Biol (2005) 170:1101-11.

[192] Sarkar S, Davies JE, Huang Z, Tunnacliffe A, Rubinztein DC. Trehalose, a novel mTOR-independent autophagy enhancer, accelerates the clearance of mutant huntingtin and alpha-synuclein. J Biol Chem (2007) 282:5641-5652.

[193] Aguib Y, Heiseke A, Gilch S, Riemer C, Baier M, Schätzl HM, Ertmer A. Autophagy induction by trehalose counteracts cellular prion infection. Autophagy (2009) 5:361-9.

[194] Sunyach C, Jen A, Deng J, Fitzgerald KT, Frobert Y, Grassi J, McCaffrey MW, Morris R. The mechanism of internalization of glycosylphosphatidylinositol-anchored prion protein. EMBO J (2003) 22:3591-3601.

[195] Marella M, Lehmann S, Grassi J, Chabry J. Filipin prevents pathological prion protein accumulation by reducing endocytosis and inducing cellular PrP release. J Biol Chem (2002) 277:25457-25464.

[196] Parkin ET, Watt NT, Turner AJ, Hooper NM. Dual mechanisms for shedding of the cellular prion protein. J Biol Chem (2004) 279:11170-11178.

[197] Aguib Y, Gilch S, Krammer C, Ertmer A, Groschup MH, Schätzl HM. Neuroendocrine cultured cells counteract persistent prion infection by downregulation of PrPc. Mol Cell Neurosci (2008) 38:98-109.

[198] Heiseke A, Schöbel S, Lichtenthaler SF, Vorberg I, Groschup MH, Kretzschmar H, Schätzl HM, Nunziante M. The novel sorting nexin SNX33 interferes with cellular PrP formation by modulation of PrP shedding. Traffic (2008) 9:1116-1129.

[199] Seo JS, Moon MH, Jeong JK, Seol JW, Lee YJ, Park BH, Park SY. SIRT1, a histone deacetylase, regulates prion protein-induced neuronal cell death. Neurobiol Aging (2012) 33:1110-1120.

[200] Jeong JK, Moon MH, Lee YJ, Seol JW, Park SY. Autophagy induced by the class III histone deacetylase Sirt1 prevents prion peptide neurotoxicity. Neurobiol Aging (2013) 34:146-156.

[201] Fornai F, Longone P, Cafaro L, Kastsiuchenka O, Ferrucci M, Manca ML, Lazzeri G, Spalloni A, Bellio N, Lenzi P, Modugno N, Siciliano G, Isidoro C, Murri L, Ruggieri S, Paparelli A. Lithium delays progression of amyotrophic lateral sclerosis. Proc Natl Acad Sci USA 105:2052-7. Erratum in: Proc Natl Acad Sci USA (2008) 105:16404-7.

[202] Ravikumar B, Vacher C, Berger Z, Davies JE, Luo S, Oroz LG, Scaravilli F, Easton DF, Duden R, O'Kane CJ, Rubinsztein DC. Inhibition of mTOR induces autophagy and reduces toxicity of polyglutamine expansions in fly and mouse models of Huntington disease. Nat Genet (2004) 36:585-95.

[203] Williams A, Jahreiss L, Sarkar S, Saiki S, Menzies FM, Ravikumar B, Rubinsztein DC. Aggregate-prone proteins are cleared from the cytosol by autophagy: therapeutic implications. Curr Top Dev Biol (2006) 76:89-101.

[204] Sakaguchi S, Katamine S, Nishida N, Moriuchi R, Shigematsu K, Sugimoto T, Nakatani A, Kataoka Y, Houtani T, Shirabe S, Okada H, Hasegawa S, Miyamoto T, Noda

T. Loss of cerebellar Purkinje cells in aged mice homozygous for a disrupted PrP gene. Nature (1996) 380:528-531.

[205] Moore RC, Redhead NJ, Selfridge J, Hope J, Manson JC, Melton DW. Double replacement gene targeting for the production of a series of mouse strains with different prion protein gene alterations. Biotechnology (1995) 13:999-1004.

[206] Rossi D, Cozzio A, Flechsig E, Klein MA, Rulicke T, Aguzzi A, Weissmann C. Onset of ataxia and Purkinje cell loss in PrP null mice inversely correlated with Dpl level in brain. EMBO J (2001) 20:694-702.

[207] Yokoyama T, Kimura KM, Ushiki Y, Yamada S, Morooka A, Nakashiba T, Sassa T, Itohara S. In vivo conversion of cellular prion protein to pathogenic isoforms, as monitored by conformation-specific antibodies. J Biol Chem (2001) 276:11265-71.

[208] Flechsig E, Hegyi I, Leimeroth R, Zuniga A, Rossi D, Cozzio A, Schwarz P, Rülicke T, Götz J, Aguzzi A, Weissmann C. Expression of truncated PrP targeted to Purkinje cells of PrP knockout mice causes Purkinje cell death and ataxia. EMBO J (2003) 22:3095-3101.

[209] Anderson L, Rossi D, Linehan J, Brandner S, Weissmann C. Transgene-driven expression of the Doppel protein in Purkinje cells causes Purkinje cell degeneration and motor impairment. Proc Natl Acad Sci USA (2004) 101:3644-3649.

[210] Yamaguchi N, Sakaguchi S, Shigematsu K, Okimura N, Katamine S. Doppel-induced Purkinje cell death is stoichiometrically abrogated by prion protein. Biochem Biophys Res Commun (2004) 319:1247-1252.

[211] Nishida N, Tremblay P, Sugimoto T, Shigematsu K, Shirabe S, Petromilli C, Erpel SP, Nakaoke R, Atarashi R, Houtani T, Torchia M, Sakaguchi S, DeArmond SJ, Prusiner SB, Katamine S. A mouse prion protein transgene rescues mice deficient for the prion protein gene from Purkinje cell degeneration and demyelination. Lab Invest (1999) 79:689-697.

[212] Cui T, Holme A, Sassoon J, Brown DR. Analysis of doppel protein toxicity. Mol Cell Neurosci (2003) 23:144-155.

[213] Atarashi R, Nishida N, Shigematsu K, Goto S, Kondo T, Sakaguchi S, Katamine S. Deletion of N-terminal residues 23-88 from prion protein (PrP) abrogates the potential to rescue PrP-deficient mice from PrP-like protein/doppel-induced neurodegeneration. J Biol Chem (2003) 278:28944-28949.

[214] Shmerling D, Hegyi I, Fischer M, Blattler T, Brandner S, Gotz J, Rulicke T, Flechsig E, Cozzio A, von Mering C, Hangartner C, Aguzzi A, Weissmann C. Expression of amino-terminally truncated PrP in the mouse leading to ataxia and specific cerebellar lesions. Cell (1998) 93:203-214.

[215] Sakudo A, Lee DC, Nakamura I, Taniuchi Y, Saeki K, Matsumoto Y, Itohara S, Ikuta K, Onodera T. Cell-autonomous PrP-Doppel interaction regulates apoptosis in PrP gene-deficient neuronal cells. Biochem Biophys Res Commun (2005) 333:448-454.

[216] Wong BS, Liu T, Paisley D, Li R, Pan T, Chen SG, Perry G, Petersen RB, Smith MA, Melton DW, Gambetti P, Brown DR, Sy MS. Induction of HO-1 and NOS in doppel-expressing mice devoid of PrP: implication for doppel function. Mol Cell Neurosci (2001) 17:768-775.

[217] Diarra-Mehrpour M, Arrabal S, Jalil A, Pinson X, Gaudin C, Pietu G, Pitaval A, Ripoche H, Eloit M, Dormont D, Chouaib S. Prion protein prevents human breast carcinoma cell line from tumor necrosis factor alpha-induced cell death. Cancer Res (2004) 64:719-727.

[218] Paitel E, Sunyach C, Alves da Costa C, Bourdon JC, Vincent B, Checler F. Primary cultured neurons devoid of cellular prion display lower responsiveness to staurosporine through the control of p53 at both transcriptional and post-transcriptional levels. J Biol Chem (2004) 279:612-618.

[219] Solforosi L, Criado JR, McGavern DB, Wirz S, Sanchez-Alavez M, Sugama S, DeGiorgio LA, Volpe BT, Wiseman E, Abalos G, Masliah E, Gilden D, Oldstone MB, Conti B, Williamson RA. Cross-linking cellular prion protein triggers neuronal apoptosis in vivo. Science (2004) 303:1514-1516.

[220] Bounhar Y, Zhang Y, Goodyer CG, LeBlanc A. Prion protein protects human neurons against Bax-mediated apoptosis. J Biol Chem (2001) 276:39145-39149.

[221] Roucou X, Guo Q, Zhang Y, Goodyer CG, LeBlanc A. Cytosolic prion protein is not toxic and protects against Bax-mediated cell death in human primary neurons. J Biol Chem (2003) 278:40877-40881.

[222] Roucou X, Giannopoulos PN, Zhang Y, Jodoin J, Goodyer CG, LeBlanc A. Cellular prion protein inhibits proapoptic Bax conformational change in human neurons and in breast carcinoma MCF-7 cells. Cell Death Differ (2005) 12:783-795.

[223] Didonna A, Sussman J, Benetti F Legname G. The role of Bax and caspase-3 in doppel-induced apoptosis of cerebellar granule cells. Prion (2012) 6:309-316.

[224] Heitz S, Lutz Y, Rodeau J-L, Zanjani H, Gautheron V, Bombarde G, Richard F, Fuchs J-P, Vogel MW, Mariani J, Bailly Y. BAX contributes to Doppel-induced apoptosis of prion-deficient Purkinje cells. Dev Neurobiol (2007) 67:670-686.

[225] Heitz S, Gautheron V, Lutz Y, Rodeau J-L, Zanjani H, Sugihara I, Bombarde G, Richard F, Fuchs J-P, Vogel MW, Mariani J, Bailly Y. BCL-2 counteracts Doppel-induced apoptosis of prion protein-deficient Purkinje cells in the Ngsk Prnp$^{0/0}$ mouse. Dev Neurobiol (2008) 68:332-348.

[226] Yin XM, Oltvai ZN, Korsmeyer SJ. BH1 and BH2 domains of Bcl-2 are required for inhibition of apoptosis and heterodimerisation with Bax. Nature (1994) 369:321-323.

[227] Roucou X, Gains M, LeBlanc A. Neuroprotective functions of prion protein. J Neurosci Res (2004) 75:153-161.

[228] Oltvai ZN, Milliman CL, Korsmeyer SJ. Bcl-2 heterodimerizes in vivo with a conserved homolg, Bax, that accelerates programmed cell death. Cell (1993) 74:609-619.

[229] Gross A, McDonnell JM, Korsmeyer SJ. BCL-2 family members and the mitochondria in apoptosis. Genes Dev (1999) 13:1899-1911.

[230] Cheng EH, Wei MC, Weiler S, Flavell RA, Mak TW, Lindsten T, Korsmeyer SJ. BCL-2, BCL-X(L) sequester BH3 domain-only molecules preventing BAX- and BAK-mediated mitochondrial apoptosis. Mol Cell (2001) 8:705-711.

[231] Li A, Barmada S, Roth K, Harris D. N-terminally deleted forms of the prion protein activate both Bax-dependent and Bax-independent neurotoxic pathways. J Neurosci (2007) 27:852-859.

[232] Chiesa R, Piccardo P, Ghetti B, Harris DA. Neurological illness in transgenic mice expressing a prion protein with an insertional mutation. Neuron (1998) 21:1339-1351.

[233] Lee DC, Sakudo A, Kim CK, Nishimura T Saeki K, Matsumoto Y, Yokoyama T, Chen SG, Itohara S, Onodera T. Fusion of doppel to octapeptide repeat and N-terminal half of hydrophobic region of prion protein confers resistance to serum deprivation. Microbiol Immunol (2006) 50:203-209.

[234] Heitz S, Grant NJ, Bailly Y. Doppel induces autophagic stress in prion protein-deficient Purkinje cells. Autophagy (2009) 5:422-424.

[235] Heitz S, Grant NJ, Leschiera R, Haeberlé A-M, Demais V, Bombarde G, Bailly Y. Autophagy and cell death of Purkinje cells overexpressing Doppel in Ngsk Prnp-deficient mice. Brain Pathol (2010) 20:119-132.

[236] Dole S, Heitz S, Bombarde G, Haeberlé A-M, Demais V, Grant NJ, Bailly Y. New insights into Doppel neurotoxicity using cerebellar organotypic cultures from prion-protein-deficient mice. Prion 2010. Medimond International Proceedings Eds. (2010) pp7-14.

[237] Metzger F, Kapfhammer JP. Protein kinase C: its role in activity-dependent Purkinje cell dendritic development and plasticity. Cerebellum (2003) 2:206-214.

[238] Falsig J, Sonati T, Herrmann US, Saban D, Li B, Arroyo K, Ballmer B, Liberski PP, Aguzzi A. Prion pathogenesis is faithfully reproduced in cerebellar organotypic slice cultures. PLoS Pathog (2012) 8:e1002985.

[239] Barbieri G, Palumbo S, Gabrusiewicz K, Azzalin A, Marchesi N, Spedito A, Biggiogera M, Sbalchiero E, Mazzini G, Miracco C, Pirtoli L, Kaminska B, Comincini S. Silencing of cellular prion protein (PrPc) expression by DNA-antisens oligonucleotides induces autophagy-dependent cell death in glioma cells. Autophagy (2011) 7:840-853.

[240] Brown DR, Wong BS, Hafiz F, Clive C, Haswell SJ, Jones IM. Normal prion protein has an activity like that of superoxide dismutase. Biochem J (1999) 345:1-5.

[241] Brown DR, Besinger A. Prion protein expression and superoxide dismutase activity. Biochem J (1998) 334:423-429.

[242] Oh JM, Choi EK, Carp RI, Kim YS. Oxidative stress impairs autophagic flux in prion protein-deficient hippocampal cells. Autophagy (2012) 8:1448-1461.

[243] Baumann F, Tolnay M, Brabeck C, Pahnke J, Kloz U, Niemann HH, Heikenwalder M, Rulicke T, Burkle A, Aguzzi A. Lethal recessive myelin toxicity of prion protein lacking its central domain. Embo J (2007) 26:538-547.

[244] Baumann F, Pahnke J, Radovanovic I, Rulicke T, Bremer J, Tolnay M, Aguzzi A. Functionally relevant domains of the prion protein identified in vivo. PLoS One (2009) 4:e6707.

[245] Lemaire-Vieille C, Bailly Y, Erlich P, Loeuillet C, Brocard J, Haeberlé A-M, Bombarde G, Rak C, Demais V, Dumestre-Pérard C, Gagnon J, Cesbron J-Y. Ataxia with cerebellar lesions in mice expressing chimeric PrP-Dpl protein. J Neurosci (2013) (in press).

Chapter 6

Role of Autophagy in Parkinson's Disease

Grace G.Y. Lim, Chengwu Zhang and
Kah-Leong Lim

1. Introduction

Parkinson's disease (PD) is a prevalent neurodegenerative movement disorder whose occurrence crosses geographic, racial and social boundaries affecting 1-2% of the population above the age of 65 (Dorsey et al., 2007). Clinically, the disease is attended by a constellation of motoric deficits that progressively worsen with age, which ultimately leads to near total immobility. Although pathological changes are distributed in the PD brain (Braak et al., 2003), the principal lesion that underlies the characteristic motor phenotype of PD patients is unequivocally the loss of dopaminergic neurons in the *substantia nigra pars compacta* (SNpc) of the midbrain. This neuronal loss results in a severe depletion of striatal dopamine (DA) and thereby an impaired nigrostriatal system that otherwise allows an individual to execute proper, coordinated movements. Accordingly, pharmacological replacement of brain DA via L-DOPA administration represents an effective symptomatic recourse for the patient (especially during the initial stages of the disease) and remains a clinical gold standard treatment for PD. However, neither L-DOPA nor any currently available therapies could slow or stop the insidious degenerative process in the PD brain. Thus, PD remains an incurable disease. Invariably, the debilitating nature and morbidity of the disease present significant healthcare, social, emotional and economic problems. As the world population rapidly ages, these problems undoubtedly would also increase. According to a recent report, more than 4 million individuals in Europe's five most and the world's ten most populous countries are currently afflicted with PD (Dorsey et al., 2007). In less than 20 years' time, the number of PD sufferers is projected to increase to close to 10 million (i.e. in 2030). This is definitely a worrying trend, and one that aptly emphasizes the urgency to develop more effective treatment modalities for the PD patient. Towards this endeavour, a better understanding of the molecular mechanism(s) that underlies the pathogenesis of PD would definitely be helpful, as the illumination of which would allow the identification and therapeutic exploitation of key molecules/events involved in the pathogenic process.

Although a subject of intense research, the etiology of PD unfortunately remains incompletely understood. However, a broad range of studies conducted over the past few decades, including epidemiological, genetic and post-mortem analysis, as well as *in vitro* and *in vivo* modelling, have contributed significantly to our understanding of the pathogenesis of the disease. In particular, the recent identification and functional characterization of several genes, including *α-synuclein, parkin, DJ-1, PINK1* and *LRRK2*, whose mutations are causative of rare familial forms of PD have provided tremendous insights into the molecular pathways underlying dopaminergic neurodegeneration (Lim and Ng, 2009; Martin et al., 2011). Collectively, these studies implicate aberrant protein and mitochondrial homeostasis as key contributors to the development of PD, with oxidative stress likely acting an important nexus between the two pathogenic events.

2. Aberrant protein homeostasis & PD

Perhaps the most glaring evidence suggesting that protein homeostasis has gone awry in the PD brain is the presence of intra-neuronal inclusions, known as Lewy Bodies (LBs), in affected regions of the diseased brain in numbers that far exceed their occasional presence in the normal brain (Lewy, 1912). These signature inclusions of PD comprise of a plethora of protein constituents that include several PD-linked gene products such as α-synuclein, parkin, DJ-1, PINK1 and LRRK2. In a recent report, Wakabayashi and colleagues have documented more than 90 components of the LB and have grouped them into 13 functional groups (Table 1) (Wakabayashi et al., 2012). Among these, α-synuclein is recognized as the major component of LB and thought to be the key initiator of LB biogenesis.

However, whether LB biogenesis represents a cytoprotective or pathogenic mechanism in PD remains debatable. Notwithstanding this, how proteins aggregate to form LB is intriguing in the first place, as the cell is endowed with several complex surveillance machineries to detect and repair faulty proteins, and also destroy those are beyond repair rapidly (Fig. 1). In this surveillance system, the chaperones (comprising of members of the heat-shock proteins) represent the first line of defense in ensuring the correct folding and refolding of proteins (Liberek et al., 2008). When a native folding state could not be attained, the chaperones will direct the misfolded protein for proteolyic removal typically by the proteasome. Proteins that are destined for proteasome-mediated degradation are usually added a chain of ubiquitin via a reaction cascade that involves the ubiquitin-activating (E1), -conjugating (E2) and -ligating (E3) enzymes, whereby successive iso-peptide linkages are formed between the terminal residue (G76) of one ubiquitin molecule and a lysine (K) residue (most commonly K48) within another. The (G76-K48) polyubiquitinated substrate is then recognized by the 26S proteasome as a target for degradation (Pickart and Cohen, 2004). It is noteworthy to mention that although the G76-K48 chain linkage is the most common form of polyubiquitin, ubiquitin self-assembly can occur at any lysine residues within the molecule (at positions 6, 11, 27, 29, 33, 48 and 63)

(Pickart, 2000; Peng et al., 2003). In addition, proteins can also be monoubiquitinated. Notably, both K63-linked polyubiquitination and monoubiquitination of proteins are not typically associated with proteasome-mediated degradation (Pickart, 2000; Peng et al., 2003).

Group	Components	Remarks
1	α-synuclein; Neurofilaments	Structural Elements
2	Agrin; 14-3-3; Synphilin-1; Tau	α-synuclein-binding proteins
3	Dorfin; GSK-3β; NUB1; Parkin; Pin1; SIAH-1	Synphilin-1-binding proteins
4	Ubiquitin; E1; UbcH7; TRAF6; TRIM9; Proteasome subunits; PA700; PA28; β-TrCP; Cullin-1; HDAC4; NEDD8; p38; p62 (Sequestosome 1); ROC1; UCHL1	UPS-related proteins
5	LC3; GABARAP; GATE-16; Glucocerebrosidase; NBR-1	Autophagosome-lysosome system
6	γ-tubulin; HDAC6; Peri-centrin	Aggresome-related proteins
7	DJ-1; CHIP; Clusterin/apolipoprotein J; DnaJB6; Heat Shock Proteins; Torsin A, SOD1 & 2; FOXO3a	Stress response-related proteins
8	CaMKII; Casein Kinase II; CDK5, G-Protein Coupled Receptor Kinase 5; LRRK2; PINK1; IκBα; NFκB; p35; phospho-lipase C-δ; Tissue Transglutaminase	Signal transduction-related proteins
9	MAP1B; MAP2; Sept4/H5	Cytoskeletal proteins
10	Cox IV; Cytochrome C; Omi/HtrA2	Mitochondria-related proteins
11	Cyclin B; Retinoblastoma Protein	Cell cycle proteins
12	Amyloid Precursor Protein; Calbindin; Choline Acetyltransferase; Chromogranin A; Synaptophysin; Synaptotagmin; Tyrosine Hydroxylase; VMAT2	Cytosolic Proteins
13	Complement Proteins; Immunoglobulin	Immune-related proteins

Table 1. Components of Lewy Body (Wakabayashi et al., 2012)

Whilst the coupling of chaperone and ubiquitin protein system (UPS) provides an efficient way for the cell to deal with protein misfolding, there are times when the capacity of these systems may be exceeded by the production of misfolded proteins (e.g. under conditions of cellular stress). In such cases, aggregation-prone proteins that failed to be degraded may be transported along microtubules in a retrograde fashion to the microtubule organizing center to form an "aggresome", a term originally coined by Johnston and Kopito more than a decade ago (Johnston et al., 1998). According to the model, aggresome formation represents a cellular response towards proteasome impairments and their localization to the juxta-nuclear region is to facilitate their capture by lysosomes and thereby their clearance by macroautophagy (hereafter referred to as autophagy). Consistent with this, aggregation-prone proteins often generate aggresome-like structures when ectopically expressed in cultured cells in the presence of proteasome inhibition (Wong et al., 2008). Moreover, several groups including ours have demonstrated that autophagy induction promotes the clearance of aggresomes whereas the reverse is true when the bulk degradation system is inhibited (Fortun et al., 2003; Iwata et al., 2005b; Opazo et al., 2008; Wong et al., 2008).

Together, the chaperone, ubiquitin-proteasome and autophagy systems thus function in synergism to effectively counterbalance the threat of protein misfolding and aggregation. Accordingly, aberrations in one or more of these systems would be expected to promote protein aggregation and inclusion body formation, as in the case of affected neurons in the PD brain where LBs occur.

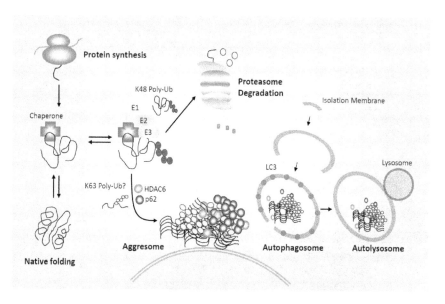

Figure 1. Schematic depiction of the collaboration among the chaperone, ubiquitin-proteasome and autophagy systems in the maintenance of intracellular protein homeostasis.

3. Biogenesis of Lewy bodies – An aggresome-related process reflecting failed autophagy?

As mentioned earlier, α-synuclein is a major component of LBs, suggesting that aberrant α-synuclein homeostasis contributes to the biogenesis of these inclusion bodies in the PD brain. The presynaptic terminal-enriched α-synuclein protein is an interesting molecule in that it is typically unfolded (or intrinsically disordered) in its native state, although the protein is extremely sensitive to its environment and can be moulded into an assortment of structurally unrelated conformations including a fibrillization-prone partially folded structure as well as various α-helical and β-sheet species occurring in both monomeric and oligomeric states (Uversky, 2007). Along with this conformation flexibility, α-synuclein also tends to misfold and becomes aggregated in the process. PD-associated mutations, including missense substitutions (A53T, A30P and E46K), duplication or triplication are known to enhance α-synuclein accumulation and aggregation (Giasson et al., 1999; Narhi et al., 1999; Conway et al., 2000; Uversky, 2007). Further, several groups have demonstrated in different experimental models that various exogenous neurotoxicants linked to PD, including pesticides, herbicides and metal ions, significantly accelerate the aggregation of α-synuclein (Manning-Bog et al., 2002; Uversky et al., 2002; Sherer et al., 2003). Not surprisingly, α-synuclein accumulation and aggregation can lead to impairments of the chaperone and UPS systems [For a recent review, see (Tan et al., 2009)]. Under such conditions, the isolation of α-synuclein aggregates into an aggresome would represent an alternative way by which the protein could be cleared, i.e. via autophagy. Indeed, emerging evidence suggest that LB biogenesis may be an aggresome-related process (Olanow et al., 2004). Because the protofibrillar, oligomeric forms of α-synuclein are thought to be more toxic than fibrillar, aggregated α-synuclein species, aggresome formation may also be regarded as a "protective" response that serves as a trap to immobilize soluble toxic forms of α-synuclein. However, this process has to be coupled to the active removal of the aggresomes by autophagy, as the unregulated growth of an inclusion body could conceivably affect cellular functions, physically or otherwise.

The relevance of aggresome formation to LB biogenesis in PD is exemplified by their striking similarities to each other in terms of structural organization, protein composition and intracellular localization (Olanow et al., 2004). For example, aggresome-related proteins such as γ-tubulin and HDAC6 can be found in LB (Table 1). HDAC6 plays an important role during aggresome formation by facilitating the retrograde transport of ubiquitinated misfolded proteins along the microtubule network to the γ-tubulin-positive MTOC by the dynein motor complex (Kawaguchi et al., 2003). Moreover, LBs are also immunopositive for p62 and NBR1, which are autophagy adapter proteins capable of binding to ubiquitinated substrates and the autophagosome protein LC3 (Bjorkoy et al., 2005; Pankiv et al., 2007; Kirkin et al., 2009). By virtue of this binding property, p62 and NBR1 may provide a link between aggresome-related proteins and their clearance by the autophagy machinery. Interestingly, all the three ubiquitin-binding autophagy receptors, i.e. p62, HDAC6 and NBR1, show preference for K63-linked polyubiquitin chains (Olzmann et al., 2007; Tan et al., 2008a; Kirkin et al., 2009), suggesting that this form of ubiquitin modification may underlie the formation as well as autophagic degradation of protein aggregates. Consistent with this, we found that K63-linked ubiquitination promotes the formation of inclusion bodies associated with PD and other neurodegenerative diseases and importantly, acts as a cargo selection signal for their subsequent removal by autophagy (Tan et al., 2008a; Tan et al., 2008b). As per our original proposal (Lim et al., 2006), it is tempting to think that the cell may switch to an alternative, proteasome-independent form of ubiquitination under conditions of proteasome-related stress that could help divert cargo proteins away from an otherwise overloaded proteasome. All these would culminate to the ultimate clearance of these proteins by autophagy (Fig. 1).

What remains curious about LB biogenesis is that it apparently takes place in the presence of constitutive autophagy, which is a characteristic of post-mitotic neurons (Wong and Cuervo, 2010). Moreover, α-synuclein is itself a substrate for autophagy (Webb et al., 2003).

Although α-synuclein can also be degraded by the proteasome, the aggregates of which appear to be preferentially cleared by the autophagy system (Petroi et al., 2012). Consistent with this, autophagy is recruited as the primary removal system in transgenic mice over-expressing oligomeric species of α-synuclein (Ebrahimi-Fakhari et al., 2011). Further, the protein can also be removed via chaperone-mediated autophagy (CMA), a specialized form of lysosomal degradation by which proteins containing a particular pentapeptide motif related to KFERQ are transported across the lysosomal membrane via the action of the integral membrane protein LAMP-2A and both cytosolic and lumenal hsc70 (Klionsky et al., 2011). Notably, the intralysosomal level of α-synuclein is significantly increased along with LAMP-2A and hsc70 in mice treated with the herbicide paraquat (which induces parkinsonism) or expressing α-synuclein as a transgene (Mak et al., 2010). Thus in theory, the level of α-synuclein, whether present as soluble or aggregated species, should be effectively managed in neurons under normal conditions or even when they are undergoing stress. Indeed, even in the PD brain, LB takes a significant length of time to develop. Given this, and that the autophagy system arguably represents the final line of cellular defense against the buildup of protein aggregates, the simplest explanation that could account for the presence of LB in PD is that the autophagy system has either become suboptimal in its function or is otherwise impaired altogether during the disease pathogenesis process.

4. Autophagy and PD

Morphological evidence of autophagic vacuole (AV) accumulation is certainly evident in PD as well as in several other neurodegenerative disorders (Anglade et al., 1997). However, whether the phenomenon represents attempts by the neuron to clean up its cobwebs of aggregated proteins, or a prelude to cell death, or simply a failure in AV consumption remains poorly understood. Notwithstanding this, two elegant studies conducted in 2006 aptly illustrated the importance of competent autophagy function to neuronal homeostasis (Hara et al., 2006; Komatsu et al., 2006). By means of targeted genetic disruption of essential components of the autophagy process (Atg5 or Atg7), these studies demonstrated that ablation of autophagy function in neural cells of mice results in extensive neurodegeneraion that is accompanied by widespread inclusion pathology, suggesting that autophagy failure can precipitate protein aggregation and subsequent cell death in affected neurons.

Supporting a role for failed autophagy in PD in the face of α-synuclein accumulation, α-synuclein was recently demonstrated to inhibit autophagy when over-expressed, both *in vitro* and *in vivo* (Winslow et al., 2010). The inhibition apparently occurs at a very early stage of autophagosome formation, which is likely a result of disrupted localization and mobilization of Atg9, a multi-spanning membrane protein whose associated vesicles are important sources of membranes for the synthesis of early autophagosomes (Yamamoto et al., 2012). Interesting, the reverse, i.e. autophagy enhancement, was observed when α-synuclein is depleted via RNAi-mediated knockdown (Winslow et al., 2010), suggesting that the protein might play a regulatory role in the synthesis of autophagosome. Moreover, targeted disruption of autophagy (via *Atg7* deletion) in midbrain dopaminergic neurons results in abnormal presynaptic accumulation of α-synuclein that is accompanied by dendritic and axonal dystrophy, reduced striatal DA content, and the formation of somatic and dendritic ubiquitinated inclusions (Friedman et al., 2012). Significant age-dependent loss of nigral dopaminergic neurons were also recorded in these *Atg7* conditionally knockout mice (*Atg7*-cKO[TH]), with 9 month old *Atg7*-cKO[TH] mice exhibiting about 40% reduction in the number of SN neurons that is accompanied by markedly decreased spontaneous motor activity and coordination relative to controls (Friedman et al., 2012). Together, these results suggest that failure in autophagy function precipitates inclusions formation in dopaminergic neurons that leads to their demise.

Besides macroautophagy, α-synuclein can also affect the function of CMA. For example, disease-associated α-synuclein mutants bind to the CMA lysosomal receptor with high affinity but are poorly translocated, resulting in the blockage of uptake and degradation of CMA substrates (Cuervo et al., 2004). The increase in cytosolic α-synuclein levels that ensued could favour its aggregation and concomitantly, amplify the burden of misfolded protein load for the cell. Interestingly, DA modification of α-synuclein also impairs CMA-mediated degradation by a similar mechanism (Martinez-Vicente et al., 2008). In this case, membrane-bound DA-α-synuclein monomers appear to seed the formation of oligomeric complexes, which consequently placed the translocation complex under siege. Consistent with this, CMA inhibition following L-DOPA treatment is more pronounced in ventral midbrain cultures containing dopaminergic neurons than in non-DA producing cortical neurons. Importantly, α-synuclein appears to be the principal mediator of DA-induced blockage of CMA, as ventral midbrain cultures derived from α-synuclein null mice are relatively spared from the inhibitory effects of DA on CMA (Martinez-Vicente et al., 2008). More recently, Malkus and Ischiropoulos demonstrated that CMA activity in the adult brain of A53T α-synuclein-expressing transgenic mice varies across different regions, with brain regions vulnerable to α-synuclein aggregation displaying marked deficiencies in CMA (Malkus and Ischiropoulos, 2012). Their results support an integral role for the lysosome in maintaining α-synuclein homeostasis and at the same time, provides an explanation to why certain brain regions are vulnerable to inclusion formation and cellular dysfunction while others are spared.

Perhaps the most direct evidence linking lysosomal dysfunction to PD is the demonstration that loss-of-function mutations in a gene encoding for the lysosomal P-type ATPase named ATP13A2 cause a juvenile and early-onset form of parkinsonism that is also characterized by pyramidal degeneration and dementia (Ramirez et al., 2006). In patient-derived fibroblasts as well as in ATP13A2-silenced primary mouse neurons, deficient ATPase function results in impaired lysosomal degradation capacity that concomitantly enhanced the accumulation and toxicity of α-synuclein (Usenovic et al., 2012). Importantly, silencing of endogenous α-synuclein ameliorated the toxicity in neurons depleted of ATP13A2, suggesting that ATP13A2-induced parkinsonism may be contributed by α-synuclein accumulation amid functional impairments of the lysosome. Supporting this, overexpression of wild type ATP13A2 suppresses α-synuclein-mediated toxicity in *C. elegans* while knockdown of ATP13A2 expression promotes the accumulation of misfolded α-synuclein in the animal (Rappley et al., 2009). Together, these studies demonstrate a functional link between ATP13A2-related lysosomal dysfunction and α-synuclein in promoting neurodegeneration.

Besides *α-synuclein* and *ATP13A2*, several other PD-linked genes have also been associated directly or indirectly with the autophagic process. For example, emerging evidence suggest that mutations in LRRK2 promote dysregulation in autophagy, although the role of LRRK2 in controlling autophagy-lysosome pathway is likely to be complex (discussed further in section 6). In the case of parkin, which has the ability to promote K63-linked ubiquitination, we and others have shown that the ubiquitin ligase is involved in aggresome formation and thereby their removal via autophagy (at least indirectly) (Lim et al., 2005; Olzmann et al., 2007). Consistent with its role as an "aggresome-promoter", parkin-related cases are frequently (although not exclusively) devoid of classic LBs, as revealed by a number of autopsy studies (Takahashi et al., 1994; Mori et al., 1998; Hayashi et al., 2000). In recent years, the attention to parkin-autophagy axis has however shifted towards its ability to remove damaged mitochondria via a specialized form of autophagy known as "mitophagy", a term originally coined by Lemasters (Lemasters, 2005). Accordingly, impairment in mitochondrial quality control due to failed mitophagy in parkin-deficient neurons is now thought to be a key mechanism that predisposes them to degeneration.

5. Mitophagy and PD

A role for mitochondria dysfunction in the pathogenesis of PD has long been appreciated. Through post-mortem analysis performed as early as 1989, several groups have recorded a significant reduction in the activity of mitochondrial complex I as well as ubiquinone (co-enzyme Q10) in the SN of PD brains (Schapira et al., 1989; Shults et al., 1997; Keeney et al., 2006). Moreover, mitochondrial poisoning recapitulates PD features in humans and represents a popular strategy to model the disease in animals (Dauer and Przedborski, 2003). Similarly, impairment of mitochondrial homeostasis via genetic ablation of TFAM, a mitochondrial transcription factor, in dopaminergic neurons of mice (named MitoPark mouse) results in energy crisis and neurodegeneration (Sterky et al., 2011).

Rather than being solitary and static structures as depicted in many textbooks, mitochondria are now recognized to be dynamic and mobile organelles that constantly undergo membrane remodeling through repeated cycles of fusion and fission as well as regulated turnover via mitophagy. These processes help to maintain a steady pool of healthy mitochondrial essential for energy production and beyond (e.g. calcium homeostasis). Following the seminal discovery by Youle group that identified parkin as a key mammalian regulator of mitophagy (Narendra et al., 2008), intensive research is now focused on elucidating the precise mechanism underlying parkin-mediated mitophagy and whether impaired clearance of damaged mitochondria may trigger the demise of dopaminergic neurons in the PD brain.

Mechanistically, the picture regarding parkin-mediated mitophagy that has emerged thus far is depicted in Figure 2.

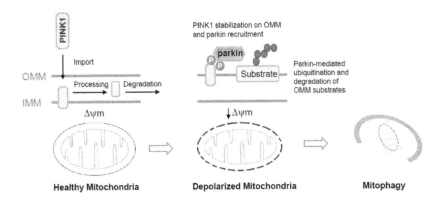

Figure 2. Model of Parkin/PINK1-mediated mitophagy

In this model, another PD-linked gene known as PINK1, which is a mitochondrial serine/ threonine kinase, collaborate closely with parkin to bring about the mitophagy process. Briefly, a key initial event that occurs upon mitochondrial depolarization is the selective accumulation of PINK1 in the outer membrane of the damaged organelle. Normally, PINK1 accumulation in healthy mitochondria is prevented by the sequential proteolytic actions of mitochondrial processing peptidase (MPP) and presenilin-associated rhomboid-like protease (PARL) that rapidly cleaves the protein to generate an unstable 53 kDa PINK1 species that is usually degraded by the proteasome or by an unknown "proteasome-like" protease (Becker et al., 2012; Greene et al., 2012). In depolarized mitochondria, PINK1 stabilization on the outer membrane enables the protein to recruit parkin to the organelle, a process that is apparently dependent on PINK1 autophosphorylation at Ser228 and Ser402 (Okatsu et al., 2012). Once recruited onto the mitochondria, parkin becomes activated and promotes the ubiquitination and subsequent degradation of many outer membrane proteins (Chan et al., 2011; Yoshii et al., 2011) including the pro-fusion mitofusin proteins (Poole et al., 2010; Ziviani et al., 2010), the elimination of which is thought to prevent unintended fusion events involving damaged mitochondria and thereby their re-entry into undamaged mitochondrial network from occurring. Mitophagy induction then occurs, which likely involves parkin-mediated K63 ubiquitination that will help recruit the autophagy adaptors HDAC6 and p62 that subsequently lead to mitochondrial clustering around the peri-nucleus region. By virtue of their association with the autophagy process, the concerted actions of p62 and HDAC6 will presumably facilitate the final removal of damaged mitochondria by the lysosome (Ding et al., 2010; Geisler et al., 2010; Lee et al., 2010). However, a recent study from Mizushima's lab revealed that the initial cargo recognition step of mitophagy does not involves the interaction between LC3 and the adaptor molecules. Rather, parkin recruitment on the mitochondria induces the formation of ULK1 (Atg1) puncta and Atg9 structures (Itakura et al., 2012). Because ULK1 complex functions as an essential upstream nucleation step of the hierachical autophagy cascade, their results suggest that mitophagosome is generated in a de novo fashion on damaged mitochondria. Autophagosomal LC3 is however important for the efficient incorporation of damaged mitochondria into the autophagosome at a later stage. Notwithstanding this, how parkin participates in the de novo synthesis of isolation membrane awaits further clarifications. Interestingly, the whole mitophagy process bears striking resemblance to the formation and autophagic clearance of aggresomes. Indeed, we have termed the mitochondrial clustering phenomenon as (formation of) "mito-aggresomes" (Lee et al., 2010). Importantly, several groups, including ours, have demonstrated that PD-associated parkin mutants are defective in supporting mitophagy due to distinct problem at recognition, transportation or ubiquitination of impaired mitochondria (Lee et al., 2010; Matsuda et al., 2010), thereby implicating dysfunctional mitophagy in the development of parkin-related parkinsonism.

Given the pivotal role of parkin/PINK1 pathway in mitochondrial quality control, it is perhaps not surprising to note that deficiency in parkin or PINK1 function results in the accumulation of abnormal mitochondria in several parkin/PINK1-related PD models. This defect is perhaps most prominently observed in *Drosophila* parkin or PINK1 mutants, especially in their flight musculature, which is plagued by pronounced mitochondrial lesions and muscle degeneration (Greene et al., 2003; Clark et al., 2006; Park et al., 2006; Wang et al., 2007). Importantly, parkin over expression in pink1-/- flies significantly ameliorates all the mutant phenotypes, although the reverse, does not happen, i.e. pink1 over expression in parkin null flies does not compensate for the loss of parkin function. These results suggest that parkin acts in the same pathway but downstream of pink1 (Clark et al., 2006; Park et al., 2006). The hierachy is consistent with the proposed model of parkin/PINK1 pathway in the regulation of mitochondrial quality control, although parkin in this case can apparently do the job in the complete absence of pink1. Notably, several other studies also suggested that mitophagy can take place in a PINK1-deficient background (Dagda et al., 2009; Cui et al., 2010; Dagda et al., 2011). Conversely, Seibler and colleagues found PINK1 to be essential for parkin-mediated mitophagy. They demonstrated that parkin recruitment to depolarized mitochondria is impaired in human dopaminergic neurons derived via the induced pluripotent stem cells route from PINK1-related PD patients, a defect that can be rescued by the re-introduction of wild-type PINK1 into PINK1-deficient neurons (Seibler et al., 2011).

As with the case with virtually all the biological models initially proposed, the parkin/PINK1 mitophagy model is currently less than perfect and clearly needs be continually updated with each new piece of significant data. The relevance of mitophagy to sporadic PD is also debatable, although we and others have previously shown that parkin dysfunction (presumably triggering mitophagy deficiency) may also underlie the pathogenesis of sporadic PD (Pawlyk et al., 2003; LaVoie et al., 2005; Wang et al., 2005a). Perhaps one of most challenging tasks at hand is to demonstrate unequivocally that mitophagy impairment, instead of a generalized impairment in the autophagy process, contributes directly to neurodegeneration *in vivo*. This would require the genetic differentiation of targeted components that are exclusively involved in mitophagy. Currently, key components of mitophagy and autophagy tend to overlap. Even parkin appear to subserve both types of autophagy processes (and more). Thus, although mitochondrial quality control is invariably important for neuronal survival, whether failure in the removal of damaged mitochondria is in itself a driver of disease pathogenesis or is a consequence of a progressive and general decline in autophagy function in the PD brain remains to be clarified.

6. Autophagy induction as therapeutic strategy for PD?

If failure in autophagy function were to underlie PD pathogenesis, it follows intuitively that stimulation of autophagy in the PD brain might be beneficial for the patient. Indeed, work from Rubinsztein lab and others have demonstrated that autophagy enhancement promotes beneficial outcomes in several experimental models of PD, supporting that such an approach could represent a viable therapeutic strategy (Rubinsztein et al., 2012).

Notably, most neurodegenerative disease-associated proteins, including α-synuclein, that are prone to aggregation are substrates of autophagy. Accordingly, pharmacological or genetic enhancement of autophagy can in theory help remove these aggregation-prone proteins and concomitantly reduce their associated toxicity. Rapamycin, an inducer of mTOR (mammalian Target of Rapamycin), is widely established to be a potent autophagy inducer. Expectedly, rapamycin treatment of cellular or animal models of α-synucleinopathies reduces the levels of both soluble and aggregated species of α-synuclein in an autophagy-dependent manner (Crews et al., 2010). Similarly, trehalose also accelerates the clearance of α-synuclein by means of its ability to induce autophagy, albeit in an mTOR-independent manner (Sarkar et al., 2007). Further, trehalose-treated cells are protected against subsequent pro-apoptotic insults. Together, trehalose and rapamycin exert an additive effect in the clearance of aggregate-prone proteins (Sarkar et al., 2007). Perhaps unsurprisingly, rapamycin can also rescue failed mitophagy in parkin deficient cells and result in improved mitochondrial function (Siddiqui et al., 2012), suggesting that generalized autophagy activation can help clean up all the cellular "cobwebs" be it protein aggregates or damaged organelles. More recently, Steele and colleagues showed that latrepirdine, a neuroactive compound associated with enhanced cognition and neuroprotection, also stimulates the degradation of α-synuclein and concomitantly protects against α-synuclein-induced toxicity in 3 model systems: yeast, differentiated SH-SY5Y cells and wild type mouse (Steele et al., 2012). The beneficial effects of latrepirdine again appear to be related to autophagy induction, as evident by the elevation of several autophagy markers in mouse brain following chronic administration of the compound. Using a genetic approach, Spencer and colleagues demonstrated via lentivirus-mediated gene transfer of beclin 1, a key promoter of autophagy, that genetic enhancement of autophagy in α-synuclein overexpressing mice ameliorates the synaptic and dendritic pathology in these transgenic animals and reduces the accumulation of the protein *in vivo* (Spencer et al., 2009). Taken together, these studies support the therapeutic applications of autophagy induction in PD, particularly in preventing the accumulation of α-synuclein.

Notwithstanding the above promising findings regarding the protective effects of autophagy induction, it is important to recognize that autophagy induction is a "double-edge sword" that can cut both ways, i.e. being protective or pro-death under different conditions. One therefore have to consider this caveat in considering autophagy induction as a therapeutic strategy for PD. Notably, the parkinsonian neurotoxin MPP+ that induces selective loss of dopaminergic neurons has been demonstrated by several groups to activate autophagy (Zhu et al., 2007; Xilouri et al., 2009; Wong et al., 2011), a process that appears to act through the dephosphorylation of LC3 (which enhances its recruitment into autophagosomes) (Cherra et al., 2010) and/or CDK5-mediated phosphorylation of endophilin B1 (which promotes its dimerization and recruitment of the UVRAG/Beclin 1 complex to induce autophagy) (Wong et al., 2011). In this case, autophagy induction is apparently harmful to dopaminergic neurons. Moreover, stimulation of autophagy also contributes to neuronal death induced by overexpression of α-synuclein (Xilouri et al., 2009). Conversely, inhibition of autophagy pharmacologically with 3-methylalanine (3-MA) or genetically via Atg5 or Atg12 gene silencing significantly attenuates neuronal loss associated with MPP + treatment or mutant α-synuclein expression, as is the case with knockdown of CDK5 or endophilin B1 (Wong et al., 2011). Along these lines, we found that mutant α-synuclein-associated toxicity is aggravated by the accumulation of iron, which act together to trigger autophagic cell death. The toxicity that α-synuclein-iron elicits can be ameliorated by pharmacological inhibition of autophagy (Chew et al., 2011). Interestingly, autophagy activation elicited by mutant α-synuclein overexpression can also result in excessive mitophagy and thereby unintended loss of mitochondria, which in turn promotes bioenergetics deficit and neuronal degeneration (Choubey et al., 2011). Further supporting a "pathological" role for autophagy, loss of DJ-1 function associated with recessive parkinsonism has been found to increase (instead of decrease) autophagic flux, although it is currently unclear how this relates to neuronal death in the context of DJ-1 deficiency (Irrcher et al., 2010).

Finally, mutations in LRRK2, which currently represent the most prevalent genetic contributor to PD, are also implicated in aberrant autophagy induction. For example, transgenic mice expressing disease-associated LRRK2 mutants (R1441C and G2019S) frequently exhibit increased incidence of autophagic vacuoles in their brain (Ramonet et al., 2011). Similarly, cells expressing G2019S LRRK2 mutant show increase autophagosome content and autophagy-dependent shortening of neurites (Plowey et al., 2008). Conversely, ablation of LRRK2 in mice promotes impairment of the autophagy pathway as evident by the accumulation of p62, lipofuscin granules, ubiquitinated proteins and α-synuclein-positive inclusions in their kidneys (Tong et al., 2010). The relationship between LRRK2 and autophagy is however complicated. For example, Gomez-Suaga and colleagues have recently demonstrated that LRRK2-induced accumulation of autophagosome is related to the ability of the kinase to activate a calcium-dependent protein kinase kinase-beta (CaMKK-beta)/adenosine monophosphate (AMP)-activated protein kinase (AMPK) pathway via modulation of NAADP-dependent Ca2+ channel on lysosomal membrane (Gomez-Suaga et al., 2012). However, they also detected at the same time a reduction in the acidification of lysosomes that can compromise autophagosome turnover and thereby autophagy (Gomez-Suaga et al., 2012), suggesting that autophagy is actually impaired rather than activated in LRRK2-expressing cells. Consistent with this, another study revealed that the expression of LRRK2 R1441C mutant leads to impaired autophagic balance that is characterized by AV accumulation containing incompletely degraded materials and increased levels of p62 (Alegre-Abarrategui et al., 2009).

Accordingly, siRNA-mediated knockdown of LRRK2 expression results in increased autophagic activity and prevented cell death caused by inhibition of autophagy in starvation conditions. Thus, the precise role of autophagy in LRRK2-related parkinsonism is anybody's guess at this moment, begging again caution in the proposed use of autophagy inducers as a therapeutic recourse.

In a related development, we recently found that disease-associated LRRK2 G2019S mutant can trigger marked mitochondrial abnormalities when overexpressed in *Drosophila*, a phenotype that can be rescued by parkin co-expression (Ng et al., 2012). Given the role of parkin in promoting mitophagy, it is tempting to speculate that the LRRK2 mutant may retard the clearance of damaged mitochondria via mitophagy in the absence of parkin overexpression. Indeed, the mitochondrial phenotype LRRK2 G2019S mutant induces in the flight muscle is reminiscent of that brought about by the loss of parkin function. Alternatively, this could also be a result of LRRK2-induced impairment in autophagy in general. Importantly, we further found that pharmacological or genetic activation of AMPK can effectively compensate for parkin deficiency to bring about a significant suppression of dopaminergic and mitochondrial dysfunction in mutant LRRK2 flies (Ng et al., 2012). Our results suggest a neuroprotective role for AMPK that might be related to mitophagy/autophagy modulation. AMPK is an evolutionarily conserved cellular energy sensor that is activated by ATP depletion or glucose starvation (Hardie, 2011). When activated, AMPK switches the cell from an anabolic to a catabolic mode and in so doing, helps to regulate diverse cellular processes that impact on cellular energy demands. Interestingly, like parkin, AMPK can also regulate mitophagy and also autophagy through its ability to phosphorylate the autophagy initiator ATG1 (Egan et al., 2011; Kim et al., 2011). Lending relevance to our findings, a recent report demonstrated that AMPK is activated in mice treated with MPTP and that inhibition of AMPK function by compound C enhances MPP(+)-induced cell death (Choi et al., 2010). More recently, a PD cohort-based study revealed that Metformin-inclusive sulfonylurea therapy reduces the risk for the disease occurring with Type 2 diabetes in a Taiwanese population (Wahlqvist et al., 2012). Metformin is a direct activator of AMPK. Together, these findings suggest that AMPK activation may protect against the development of PD, presumably via its ability to maintain energy balance via the modulation of autophagy as well as a range of other cellular processes. Given that caveats associated with direct autophagy induction, and that excessive autophagy can result in energy crisis especially in the aged brain, perhaps AMPK activation, through its ability to maintain both protein and energy homeostasis, would represent a better approach than direct autophagy induction as a therapeutic strategy for PD.

7. Conclusion

In essence, the case for autophagy dysfunction as a contributor to the pathogenesis of PD is rather compelling. As evident from the above discussion, virtually all the major PD-associated gene products have some direct or indirect relationship with the autophagy-lysosome axis. What is less clear is whether autophagy induction is neuroprotective or is a key driver of neurodegeneration. One can envisage that the activation of autophagy may be beneficial in the short term (particularly when the induction is transient and timely), but deleterious when it is becomes chronic or excessive. Finding the tipping autophagy threshold point between neuroprotection and neurodegeneration would therefore be an important endeavour, the clarification of which has important implications for the future development of autophagy-related therapeutics for the PD patient.

Acknowledgements

This work was supported by grants from Singapore Millennium Foundation, National Medical Research Council and A*STAR Biomedical Research Council (LKL). G.L. is supported by a graduate scholarship from the Singapore Millennium Foundation.

Author details

Grace G.Y. Lim[1], Chengwu Zhang[2] and Kah-Leong Lim[1,2,3]

1 Department of Physiology, National University of Singapore, Singapore

2 National Neuroscience Institute, Singapore

3 Duke-NUS Graduate Medical School, Singapore

References

[1] Alegre-abarrategui, J, Christian, H, Lufino, M. M, Mutihac, R, Venda, L. L, Ansorge, O, & Wade-martins, R. (2009). LRRK2 regulates autophagic activity and localizes to specific membrane microdomains in a novel human genomic reporter cellular model. Hum Mol Genet , 18, 4022-4034.

[2] Anglade, P, Vyas, S, Javoy-agid, F, Herrero, M. T, Michel, P. P, Marquez, J, Mouatt-prigent, A, Ruberg, M, Hirsch, E. C, & Agid, Y. (1997). Apoptosis and autophagy in nigral neurons of patients with Parkinson's disease. Histol Histopathol , 12, 25-31.

[3] Becker, D, Richter, J, Tocilescu, M. A, Przedborski, S, & Voos, W. (2012). Pink1 kinase and its membrane potential (Deltapsi)-dependent cleavage product both localize to outer mitochondrial membrane by unique targeting mode. J Biol Chem , 287, 22969-22987.

[4] Bjorkoy, G, Lamark, T, Brech, A, Outzen, H, Perander, M, Overvatn, A, Stenmark, H, & Johansen, T. forms protein aggregates degraded by autophagy and has a protective effect on huntingtin-induced cell death. J Cell Biol , 171, 603-614.

[5] Braak, H. Del Tredici K, Rub U, de Vos RA, Jansen Steur EN, Braak E ((2003). Staging of brain pathology related to sporadic Parkinson's disease. Neurobiol Aging , 24, 197-211.

[6] Chan, N. C, Salazar, A. M, Pham, A. H, Sweredoski, M. J, Kolawa, N. J, Graham, R. L, Hess, S, & Chan, D. C. (2011). Broad activation of the ubiquitin-proteasome system by Parkin is critical for mitophagy. Hum Mol Genet , 20, 1726-1737.

[7] Cherra, S. J. rd, Kulich SM, Uechi G, Balasubramani M, Mountzouris J, Day BW, Chu CT ((2010). Regulation of the autophagy protein LC3 by phosphorylation. J Cell Biol , 190, 533-539.

[8] Chew, K. C, Ang, E. T, Tai, Y. K, Tsang, F, Lo, S. Q, Ong, E, Ong, W. Y, Shen, H. M, Lim, K. L, Dawson, V. L, Dawson, T. M, & Soong, T. W. (2011). Enhanced autophagy from chronic toxicity of iron and mutant A53T alpha-synuclein: implications for neuronal cell death in Parkinson disease. J Biol Chem , 286, 33380-33389.

[9] Choi, J. S, Park, C, & Jeong, J. W. (2010). AMP-activated protein kinase is activated in Parkinson's disease models mediated by 1-methyl-4-phenyl-1,2,3,6-tetrahydropyridine. Biochem Biophys Res Commun , 391, 147-151.

[10] Choubey, V, Safiulina, D, Vaarmann, A, Cagalinec, M, Wareski, P, Kuum, M, Zharkovsky, A, & Kaasik, A. induces neuronal death by increasing mitochondrial autophagy. J Biol Chem , 286, 10814-10824.

[11] Clark, I. E, Dodson, M. W, Jiang, C, Cao, J. H, Huh, J. R, Seol, J. H, Yoo, S. J, Hay, B. A, & Guo, M. (2006). Drosophila pink1 is required for mitochondrial function and interacts genetically with parkin. Nature , 441, 1162-1166.

[12] Conway, K. A, Lee, S. J, Rochet, J. C, Ding, T. T, Williamson, R. E, & Lansbury, P. T. Jr. ((2000). Acceleration of oligomerization, not fibrillization, is a shared property of both alpha-synuclein mutations linked to early-onset Parkinson's disease: implications for pathogenesis and therapy. Proc Natl Acad Sci U S A , 97, 571-576.

[13] Crews, L, Spencer, B, Desplats, P, Patrick, C, Paulino, A, Rockenstein, E, Hansen, L, Adame, A, Galasko, D, & Masliah, E. (2010). Selective molecular alterations in the autophagy pathway in patients with Lewy body disease and in models of alpha-synucleinopathy. PLoS One 5:e9313.

[14] Cuervo, A. M, Stefanis, L, Fredenburg, R, Lansbury, P. T, & Sulzer, D. (2004). Impaired degradation of mutant alpha-synuclein by chaperone-mediated autophagy. Science , 305, 1292-1295.

[15] Cui, M, Tang, X, Christian, W. V, Yoon, Y, & Tieu, K. (2010). Perturbations in mitochondrial dynamics induced by human mutant PINK1 can be rescued by the mitochondrial division inhibitor mdivi-1. J Biol Chem , 285, 11740-11752.

[16] Dagda, R. K, & Cherra, S. J. rd, Kulich SM, Tandon A, Park D, Chu CT ((2009). Loss of PINK1 function promotes mitophagy through effects on oxidative stress and mitochondrial fission. J Biol Chem , 284, 13843-13855.

[17] Dagda, R. K, Gusdon, A. M, Pien, I, Strack, S, Green, S, Li, C, Van Houten, B, & Cherra, S. J. rd, Chu CT ((2011). Mitochondrially localized PKA reverses mitochondrial pathology and dysfunction in a cellular model of Parkinson's disease. Cell Death Differ , 18, 1914-1923.

[18] Dauer, W, & Przedborski, S. (2003). Parkinson's disease: mechanisms and models. Neuron , 39, 889-909.

[19] Ding, W. X, Ni, H. M, Li, M, Liao, Y, Chen, X, Stolz, D. B, & Dorn, G. W. nd, Yin XM ((2010). Nix is critical to two distinct phases of mitophagy, reactive oxygen species-mediated autophagy induction and Parkin-ubiquitin-mitochondrial priming. J Biol Chem 285:27879-27890., 62.

[20] Dorsey, E. R, Constantinescu, R, Thompson, J. P, Biglan, K. M, Holloway, R. G, Kieburtz, K, Marshall, F. J, Ravina, B. M, Schifitto, G, Siderowf, A, & Tanner, C. M. (2007). Projected number of people with Parkinson disease in the most populous nations, 2005 through 2030. Neurology , 68, 384-386.

[21] Egan, D. F, Shackelford, D. B, Mihaylova, M. M, Gelino, S, Kohnz, R. A, Mair, W, Vasquez, D. S, Joshi, A, Gwinn, D. M, Taylor, R, Asara, J. M, Fitzpatrick, J, Dillin, A, Viollet, B, Kundu, M, Hansen, M, & Shaw, R. J. (2011). Phosphorylation of ULK1 (hATG1) by AMP-activated protein kinase connects energy sensing to mitophagy. Science , 331, 456-461.

[22] Fortun, J, & Dunn, W. A. Jr., Joy S, Li J, Notterpek L ((2003). Emerging role for autophagy in the removal of aggresomes in Schwann cells. J Neurosci , 23, 10672-10680.

[23] Friedman, L. G, Lachenmayer, M. L, Wang, J, He, L, Poulose, S. M, Komatsu, M, Holstein, G. R, & Yue, Z. (2012). Disrupted autophagy leads to dopaminergic axon and dendrite degeneration and promotes presynaptic accumulation of alpha-synuclein and LRRK2 in the brain. J Neurosci , 32, 7585-7593.

[24] Geisler, S, Holmstrom, K. M, Skujat, D, Fiesel, F. C, Rothfuss, O. C, Kahle, P. J, & Springer, W. (2010). PINK1/Parkin-mediated mitophagy is dependent on VDAC1 and SQSTM1. Nat Cell Biol 12:119-131., 62.

[25] Giasson, B. I, Uryu, K, Trojanowski, J. Q, & Lee, V. M. (1999). Mutant and wild type human alpha-synucleins assemble into elongated filaments with distinct morphologies in vitro. J Biol Chem , 274, 7619-7622.

[26] Gomez-suaga, P, Luzon-toro, B, Churamani, D, Zhang, L, Bloor-young, D, Patel, S, Woodman, P. G, Churchill, G. C, & Hilfiker, S. (2012). Leucine-rich repeat kinase 2 regulates autophagy through a calcium-dependent pathway involving NAADP. Hum Mol Genet , 21, 511-525.

[27] Greene, A. W, Grenier, K, Aguileta, M. A, Muise, S, Farazifard, R, Haque, M. E, Mcbride, H. M, & Park, D. S. Fon EA ((2012). Mitochondrial processing peptidase regulates PINK1 processing, import and Parkin recruitment. EMBO Rep , 13, 378-385.

[28] Greene, J. C, Whitworth, A. J, Kuo, I, Andrews, L. A, Feany, M. B, & Pallanck, L. J. (2003). Mitochondrial pathology and apoptotic muscle degeneration in Drosophila parkin mutants. Proc Natl Acad Sci U S A , 100, 4078-4083.

[29] Hara, T, Nakamura, K, Matsui, M, Yamamoto, A, Nakahara, Y, Suzuki-migishima, R, Yokoyama, M, Mishima, K, Saito, I, Okano, H, & Mizushima, N. (2006). Suppression of basal autophagy in neural cells causes neurodegenerative disease in mice. Nature , 441, 885-889.

[30] Hardie, D. G. (2011). AMP-activated protein kinase: an energy sensor that regulates all aspects of cell function. Genes Dev , 25, 1895-1908.

[31] Hayashi, S, Wakabayashi, K, Ishikawa, A, Nagai, H, Saito, M, Maruyama, M, Takahashi, T, Ozawa, T, Tsuji, S, & Takahashi, H. (2000). An autopsy case of autosomal-recessive juvenile parkinsonism with a homozygous exon 4 deletion in the parkin gene. Mov Disord , 15, 884-888.

[32] Irrcher, I, et al. (2010). Loss of the Parkinson's disease-linked gene DJ-1 perturbs mitochondrial dynamics. Hum Mol Genet , 19, 3734-3746.

[33] Itakura, E, Kishi-itakura, C, Koyama-honda, I, & Mizushima, N. (2012). Structures containing Atg9A and the ULK1 complex independently target depolarized mitochondria at initial stages of Parkin-mediated mitophagy. J Cell Sci , 125, 1488-1499.

[34] Iwata, A, Riley, B. E, Johnston, J. A, & Kopito, R. R. and microtubules are required for autophagic degradation of aggregated huntingtin. J Biol Chem , 280, 40282-40292.

[35] Johnston, J. A, Ward, C. L, & Kopito, R. R. (1998). Aggresomes: a cellular response to misfolded proteins. J Cell Biol , 143, 1883-1898.

[36] Kawaguchi, Y, Kovacs, J. J, Mclaurin, A, Vance, J. M, Ito, A, & Yao, T. P. (2003). The deacetylase HDAC6 regulates aggresome formation and cell viability in response to misfolded protein stress. Cell , 115, 727-738.

[37] Keeney, P. M, Xie, J, Capaldi, R. A, & Bennett, J. P. Jr. ((2006). Parkinson's disease brain mitochondrial complex I has oxidatively damaged subunits and is functionally impaired and misassembled. J Neurosci , 26, 5256-5264.

[38] Kim, J, Kundu, M, Viollet, B, & Guan, K. L. (2011). AMPK and mTOR regulate autophagy through direct phosphorylation of Ulk1. Nat Cell Biol , 13, 132-141.

[39] Kirkin, V, Lamark, T, Sou, Y. S, Bjorkoy, G, Nunn, J. L, Bruun, J. A, Shvets, E, Mcewan, D. G, Clausen, T. H, Wild, P, Bilusic, I, Theurillat, J. P, Overvatn, A, Ishii, T, Elazar, Z, Komatsu, M, Dikic, I, & Johansen, T. (2009). A role for NBR1 in autophagosomal degradation of ubiquitinated substrates. Mol Cell , 33, 505-516.

[40] Klionsky, D. J, et al. (2011). A comprehensive glossary of autophagy-related mole-
      cules and processes (2nd edition). Autophagy , 7, 1273-1294.

[41] Komatsu, M, Waguri, S, Chiba, T, Murata, S, Iwata, J, Tanida, I, Ueno, T, Koike, M,
      Uchiyama, Y, Kominami, E, & Tanaka, K. (2006). Loss of autophagy in the central
      nervous system causes neurodegeneration in mice. Nature , 441, 880-884.

[42] LaVoie MJOstaszewski BL, Weihofen A, Schlossmacher MG, Selkoe DJ ((2005). Dopa-
      mine covalently modifies and functionally inactivates parkin. Nat Med , 11,
      1214-1221.

[43] Lee, J. Y, Nagano, Y, Taylor, J. P, Lim, K. L, & Yao, T. P. (2010). Disease-causing mu-
      tations in Parkin impair mitochondrial ubiquitination, aggregation, and HDAC6-de-
      pendent mitophagy. J Cell Biol , 189, 671-679.

[44] Lemasters, J. J. (2005). Selective mitochondrial autophagy, or mitophagy, as a target-
      ed defense against oxidative stress, mitochondrial dysfunction, and aging. Rejuvena-
      tion Res , 8, 3-5.

[45] Lewy, F. H. (1912). Paralysis agitans. I. Pathologische Anatomie Ed. M Lewandow-
      ski:, 920-933.

[46] Liberek, K, Lewandowska, A, & Zietkiewicz, S. (2008). Chaperones in control of pro-
      tein disaggregation. Embo J , 27, 328-335.

[47] Lim, K. L, & Ng, C. H. (2009). Genetic models of Parkinson disease. Biochim Biophys
      Acta , 1792, 604-615.

[48] Lim, K. L, Dawson, V. L, & Dawson, T. M. (2006). Parkin-mediated lysine 63-linked
      polyubiquitination: a link to protein inclusions formation in Parkinson's and other
      conformational diseases? Neurobiol Aging , 27, 524-529.

[49] Lim, K. L, Chew, K. C, Tan, J. M, Wang, C, Chung, K. K, Zhang, Y, Tanaka, Y, Smith,
      W, Engelender, S, Ross, C. A, Dawson, V. L, & Dawson, T. M. (2005). Parkin mediates
      nonclassical, proteasomal-independent ubiquitination of synphilin-1: implications
      for Lewy body formation. J Neurosci , 25, 2002-2009.

[50] Mak, S. K, Mccormack, A. L, Manning-bog, A. B, & Cuervo, A. M. Di Monte DA
      ((2010). Lysosomal degradation of alpha-synuclein in vivo. J Biol Chem , 285,
      13621-13629.

[51] Malkus, K. A, & Ischiropoulos, H. (2012). Regional deficiencies in chaperone-mediat-
      ed autophagy underlie alpha-synuclein aggregation and neurodegeneration. Neuro-
      biol Dis , 46, 732-744.

[52] Manning-bog, A. B, Mccormack, A. L, Li, J, Uversky, V. N, & Fink, A. L. Di Monte
      DA ((2002). The herbicide paraquat causes up-regulation and aggregation of alpha-
      synuclein in mice: paraquat and alpha-synuclein. J Biol Chem , 277, 1641-1644.

[53] Martin, I, Dawson, V. L, & Dawson, T. M. (2011). Recent advances in the genetics of Parkinson's disease. Annu Rev Genomics Hum Genet , 12, 301-325.

[54] Martinez-vicente, M, Talloczy, Z, Kaushik, S, Massey, A. C, Mazzulli, J, Mosharov, E. V, Hodara, R, Fredenburg, R, Wu, D. C, Follenzi, A, Dauer, W, Przedborski, S, Ischiropoulos, H, Lansbury, P. T, Sulzer, D, & Cuervo, A. M. (2008). Dopamine-modified alpha-synuclein blocks chaperone-mediated autophagy. J Clin Invest , 118, 777-788.

[55] Matsuda, N, Sato, S, Shiba, K, Okatsu, K, Saisho, K, Gautier, C. A, Sou, Y. S, Saiki, S, Kawajiri, S, Sato, F, Kimura, M, Komatsu, M, Hattori, N, & Tanaka, K. (2010). PINK1 stabilized by mitochondrial depolarization recruits Parkin to damaged mitochondria and activates latent Parkin for mitophagy. J Cell Biol , 189, 211-221.

[56] Mori, H, Kondo, T, Yokochi, M, Matsumine, H, Nakagawa-hattori, Y, Miyake, T, Suda, K, & Mizuno, Y. (1998). Pathologic and biochemical studies of juvenile parkinsonism linked to chromosome 6q. Neurology , 51, 890-892.

[57] Narendra, D, Tanaka, A, Suen, D. F, & Youle, R. J. (2008). Parkin is recruited selectively to impaired mitochondria and promotes their autophagy. J Cell Biol , 183, 795-803.

[58] Narhi, L, Wood, S. J, Steavenson, S, Jiang, Y, Wu, G. M, Anafi, D, Kaufman, S. A, Martin, F, Sitney, K, Denis, P, Louis, J. C, Wypych, J, Biere, A. L, & Citron, M. (1999). Both familial Parkinson's disease mutations accelerate alpha-synuclein aggregation. J Biol Chem , 274, 9843-9846.

[59] Ng, C. H, Guan, M. S, Koh, C, Ouyang, X, Yu, F, Tan, E. K, Neill, O, Zhang, S. P, Chung, X, & Lim, J. KL ((2012). AMP Kinase Activation Mitigates Dopaminergic Dysfunction and Mitochondrial Abnormalities in Drosophila Models of Parkinson's Disease. J Neurosci , 32, 14311-14317.

[60] Okatsu, K, Oka, T, Iguchi, M, Imamura, K, Kosako, H, Tani, N, Kimura, M, Go, E, Koyano, F, Funayama, M, Shiba-fukushima, K, Sato, S, Shimizu, H, Fukunaga, Y, Taniguchi, H, Komatsu, M, Hattori, N, Mihara, K, Tanaka, K, & Matsuda, N. (2012). PINK1 autophosphorylation upon membrane potential dissipation is essential for Parkin recruitment to damaged mitochondria. Nat Commun 3:1016.

[61] Olanow, C. W, Perl, D. P, Demartino, G. N, & Mcnaught, K. S. (2004). Lewy-body formation is an aggresome-related process: a hypothesis. Lancet Neurol , 3, 496-503.

[62] Olzmann, J. A, Li, L, Chudaev, M. V, Chen, J, Perez, F. A, Palmiter, R. D, & Chin, L. S. linked polyubiquitination targets misfolded DJ-1 to aggresomes via binding to HDAC6. J Cell Biol , 178, 1025-1038.

[63] Opazo, F, Krenz, A, Heermann, S, Schulz, J. B, & Falkenburger, B. H. (2008). Accumulation and clearance of alpha-synuclein aggregates demonstrated by time-lapse imaging. J Neurochem , 106, 529-540.

[64] Pankiv, S, Clausen, T. H, Lamark, T, Brech, A, Bruun, J. A, Outzen, H, Overvatn, A, Bjorkoy, G, & Johansen, T. Binds Directly to Atg8/LC3 to Facilitate Degradation of Ubiquitinated Protein Aggregates by Autophagy. J Biol Chem , 282, 24131-24145.

[65] Park, J, Lee, S. B, Lee, S, Kim, Y, Song, S, Kim, S, Bae, E, Kim, J, Shong, M, Kim, J. M, & Chung, J. (2006). Mitochondrial dysfunction in Drosophila PINK1 mutants is complemented by parkin. Nature , 441, 1157-1161.

[66] Pawlyk, A. C, Giasson, B. I, Sampathu, D. M, Perez, F. A, Lim, K. L, Dawson, V. L, Dawson, T. M, Palmiter, R. D, Trojanowski, J. Q, & Lee, V. M. (2003). Novel monoclonal antibodies demonstrate biochemical variation of brain parkin with age. J Biol Chem , 278, 48120-48128.

[67] Peng, J, Schwartz, D, Elias, J. E, Thoreen, C. C, Cheng, D, Marsischky, G, Roelofs, J, Finley, D, & Gygi, S. P. (2003). A proteomics approach to understanding protein ubiquitination. Nat Biotechnol , 21, 921-926.

[68] Pickart, C. M. (2000). Ubiquitin in chains. Trends Biochem Sci , 25, 544-548.

[69] Pickart, C. M, & Cohen, R. E. (2004). Proteasomes and their kin: proteases in the machine age. Nat Rev Mol Cell Biol , 5, 177-187.

[70] Plowey, E. D, & Cherra, S. J. rd, Liu YJ, Chu CT ((2008). Role of autophagy in G2019S-LRRK2-associated neurite shortening in differentiated SH-SY5Y cells. J Neurochem , 105, 1048-1056.

[71] Poole, A. C, Thomas, R. E, Yu, S, Vincow, E. S, & Pallanck, L. (2010). The mitochondrial fusion-promoting factor mitofusin is a substrate of the PINK1/parkin pathway. PLoS One 5:e10054.

[72] Ramirez, A, Heimbach, A, Grundemann, J, Stiller, B, Hampshire, D, Cid, L. P, Goebel, I, Mubaidin, A. F, Wriekat, A. L, Roeper, J, Al-din, A, Hillmer, A. M, Karsak, M, Liss, B, Woods, C. G, Behrens, M. I, & Kubisch, C. (2006). Hereditary parkinsonism with dementia is caused by mutations in ATP13A2, encoding a lysosomal type 5 P-type ATPase. Nat Genet , 38, 1184-1191.

[73] Ramonet, D, et al. (2011). Dopaminergic neuronal loss, reduced neurite complexity and autophagic abnormalities in transgenic mice expressing G2019S mutant LRRK2. PLoS One 6:e18568.

[74] Rappley, I, Gitler, A. D, & Selvy, P. E. LaVoie MJ, Levy BD, Brown HA, Lindquist S, Selkoe DJ ((2009). Evidence that alpha-synuclein does not inhibit phospholipase D. Biochemistry , 48, 1077-1083.

[75] Rubinsztein, D. C, Codogno, P, & Levine, B. (2012). Autophagy modulation as a potential therapeutic target for diverse diseases. Nat Rev Drug Discov , 11, 709-730.

[76] Sarkar, S, Davies, J. E, Huang, Z, Tunnacliffe, A, & Rubinsztein, D. C. (2007). Trehalose, a novel mTOR-independent autophagy enhancer, accelerates the clearance of mutant huntingtin and alpha-synuclein. J Biol Chem , 282, 5641-5652.

[77] Schapira, A. H, Cooper, J. M, Dexter, D, Jenner, P, Clark, J. B, & Marsden, C. D. (1989). Mitochondrial complex I deficiency in Parkinson's disease. Lancet 1:1269.

[78] Seibler, P, Graziotto, J, Jeong, H, Simunovic, F, Klein, C, & Krainc, D. (2011). Mitochondrial Parkin Recruitment Is Impaired in Neurons Derived from Mutant PINK1 Induced Pluripotent Stem Cells. J Neurosci , 31, 5970-5976.

[79] Sherer, T. B, Kim, J. H, Betarbet, R, & Greenamyre, J. T. (2003). Subcutaneous rotenone exposure causes highly selective dopaminergic degeneration and alpha-synuclein aggregation. Exp Neurol , 179, 9-16.

[80] Shults, C. W, Haas, R. H, Passov, D, & Beal, M. F. levels correlate with the activities of complexes I and II/III in mitochondria from parkinsonian and nonparkinsonian subjects. Ann Neurol , 42, 261-264.

[81] Siddiqui, A, Hanson, I, & Andersen, J. K. (2012). Mao-B elevation decreases parkin's ability to efficiently clear damaged mitochondria: protective effects of rapamycin. Free Radic Res , 46, 1011-1018.

[82] Spencer, B, Potkar, R, Trejo, M, Rockenstein, E, Patrick, C, Gindi, R, Adame, A, Wysscoray, T, & Masliah, E. (2009). Beclin 1 gene transfer activates autophagy and ameliorates the neurodegenerative pathology in alpha-synuclein models of Parkinson's and Lewy body diseases. J Neurosci , 29, 13578-13588.

[83] Steele, J. W, et al. (2012). Latrepirdine stimulates autophagy and reduces accumulation of alpha-synuclein in cells and in mouse brain. Mol Psychiatry.

[84] Sterky, F. H, Lee, S, Wibom, R, Olson, L, & Larsson, N. G. (2011). Impaired mitochondrial transport and Parkin-independent degeneration of respiratory chain-deficient dopamine neurons in vivo. Proc Natl Acad Sci U S A , 108, 12937-12942.

[85] Takahashi, H, Ohama, E, Suzuki, S, Horikawa, Y, Ishikawa, A, Morita, T, Tsuji, S, & Ikuta, F. (1994). Familial juvenile parkinsonism: clinical and pathologic study in a family. Neurology , 44, 437-441.

[86] Tan, J. M, Wong, E. S, & Lim, K. L. (2009). Protein misfolding and aggregation in Parkinson's disease. Antioxid Redox Signal , 11, 2119-2134.

[87] Tan, J. M, Wong, E. S, Dawson, V. L, Dawson, T. M, & Lim, K. L. linked polyubiquitin potentially partners with to promote the clearance of protein inclusions by autophagy. Autophagy 4:251-253., 62.

[88] Tan, J. M, Wong, E. S, Kirkpatrick, D. S, Pletnikova, O, Ko, H. S, Tay, S. P, Ho, M. W, Troncoso, J, Gygi, S. P, Lee, M. K, Dawson, V. L, Dawson, T. M, & Lim, K. L. linked

ubiquitination promotes the formation and autophagic clearance of protein inclusions associated with neurodegenerative diseases. Hum Mol Genet , 17, 431-439.

[89] Tong, Y, Yamaguchi, H, Giaime, E, Boyle, S, Kopan, R, & Kelleher, R. J. rd, Shen J ((2010). Loss of leucine-rich repeat kinase 2 causes impairment of protein degradation pathways, accumulation of alpha-synuclein, and apoptotic cell death in aged mice. Proc Natl Acad Sci U S A , 107, 9879-9884.

[90] Usenovic, M, Tresse, E, Mazzulli, J. R, Taylor, J. P, & Krainc, D. (2012). Deficiency of ATP13A2 leads to lysosomal dysfunction, alpha-synuclein accumulation, and neurotoxicity. J Neurosci , 32, 4240-4246.

[91] Uversky, V. N. (2007). Neuropathology, biochemistry, and biophysics of alpha-synuclein aggregation. J Neurochem , 103, 17-37.

[92] Uversky, V. N, Li, J, Bower, K, & Fink, A. L. (2002). Synergistic effects of pesticides and metals on the fibrillation of alpha-synuclein: implications for Parkinson's disease. Neurotoxicology , 23, 527-536.

[93] Wahlqvist, M. L, Lee, M. S, Hsu, C. C, Chuang, S. Y, Lee, J. T, & Tsai, H. N. (2012). Metformin-inclusive sulfonylurea therapy reduces the risk of Parkinson's disease occurring with Type 2 diabetes in a Taiwanese population cohort. Parkinsonism Relat Disord.

[94] Wakabayashi, K, Tanji, K, Odagiri, S, Miki, Y, Mori, F, & Takahashi, H. (2012). The Lewy Body in Parkinson's Disease and Related Neurodegenerative Disorders. Mol Neurobiol.

[95] Wang, C, Lu, R, Ouyang, X, Ho, M. W, Chia, W, Yu, F, & Lim, K. L. (2007). Drosophila overexpressing parkin R275W mutant exhibits dopaminergic neuron degeneration and mitochondrial abnormalities. J Neurosci , 27, 8563-8570.

[96] Wang, C, Ko, H. S, Thomas, B, Tsang, F, Chew, K. C, Tay, S. P, Ho, M. W, Lim, T. M, Soong, T. W, Pletnikova, O, Troncoso, J, Dawson, V. L, Dawson, T. M, & Lim, K. L. alterations in parkin solubility promote parkin aggregation and compromise parkin's protective function. Hum Mol Genet , 14, 3885-3897.

[97] Webb, J. L, Ravikumar, B, Atkins, J, Skepper, J. N, & Rubinsztein, D. C. (2003). Alpha-Synuclein is degraded by both autophagy and the proteasome. J Biol Chem , 278, 25009-25013.

[98] Winslow, A. R, Chen, C. W, Corrochano, S, Acevedo-arozena, A, Gordon, D. E, Peden, A. A, Lichtenberg, M, Menzies, F. M, Ravikumar, B, Imarisio, S, Brown, S, Kane, O, Rubinsztein, C. J, & Alpha-synuclein, D. C. impairs macroautophagy: implications for Parkinson's disease. J Cell Biol , 190, 1023-1037.

[99] Wong, A. S, Lee, R. H, Cheung, A. Y, Yeung, P. K, Chung, S. K, Cheung, Z. H, & Ip, N. Y. (2011). Cdk5-mediated phosphorylation of endophilin B1 is required for induced autophagy in models of Parkinson's disease. Nat Cell Biol , 13, 568-579.

[100] Wong, E, & Cuervo, A. M. (2010). Autophagy gone awry in neurodegenerative diseases. Nat Neurosci , 13, 805-811.

[101] Wong, E. S, Tan, J. M, Soong, W. E, Hussein, K, Nukina, N, Dawson, V. L, Dawson, T. M, Cuervo, A. M, & Lim, K. L. (2008). Autophagy-mediated clearance of aggresomes is not a universal phenomenon. Hum Mol Genet , 17, 2570-2582.

[102] Xilouri, M, Vogiatzi, T, Vekrellis, K, Park, D, & Stefanis, L. (2009). Abberant alpha-synuclein confers toxicity to neurons in part through inhibition of chaperone-mediated autophagy. PLoS One 4:e5515.

[103] Yamamoto, H, Kakuta, S, Watanabe, T. M, Kitamura, A, Sekito, T, Kondo-kakuta, C, Ichikawa, R, Kinjo, M, & Ohsumi, Y. (2012). Atg9 vesicles are an important membrane source during early steps of autophagosome formation. J Cell Biol , 198, 219-233.

[104] Yoshii, S. R, Kishi, C, Ishihara, N, & Mizushima, N. (2011). Parkin mediates proteasome-dependent protein degradation and rupture of the outer mitochondrial membrane. J Biol Chem.

[105] Zhu, J. H, Horbinski, C, Guo, F, Watkins, S, Uchiyama, Y, & Chu, C. T. (2007). Regulation of autophagy by extracellular signal-regulated protein kinases during 1-methyl-4-phenylpyridinium-induced cell death. Am J Pathol , 170, 75-86.

[106] Ziviani, E, Tao, R. N, & Whitworth, A. J. (2010). Drosophila parkin requires PINK1 for mitochondrial translocation and ubiquitinates mitofusin. Proc Natl Acad Sci U S A , 107, 5018-5023.

Autophagy and Cell Death

Chapter 7

Autophagy in Development and Remodelling of Mammary Gland

Malgorzata Gajewska, Katarzyna Zielniok and
Tomasz Motyl

Additional information is available at the end of the chapter

1. Introduction

Mammary gland is a unique organ, which undergoes the majority of its development in the postnatal life of mammals, especially during puberty and pregnancy. It is a complex tissue comprised of many cell types. The actual glandular part is formed by two epithelial cell subtypes: outer myoepithelia and inner luminal epithelia, which form a complicated net of ducts and lobules involved in milk synthesis and secretion during lactation. The glandular epithelium is embedded in stroma composed of mesenchymal cells, such as fibroblasts, adipocytes, immune cells, and extracellular matrix (ECM). The general features of mammary gland development is universal for animals and humans, however some differences in growth rate and hormonal control of the process can be distinguished between species.

During embryogenesis emergence of epithelial buds from ectoderm into mammary mesenchyme initiates formation of a rudimentary system of ducts, which continue their moderate elongation after birth, simultaneously with the increases in the body weight. In rodents prepubertal mammary gland consist of long, infrequently branching ducts terminated by highly proliferative structures, called terminal end buds (TEBs). TEBs contain two distinct cell types: cap cells organized as a single layer at the leading edge of these structures, staying in direct contact with the thin layer of basal lamina, and body cells, which form the multicellular bulk of the TEB [1] (Figure 1). In ruminants mammary ductal network develops as compact, highly branched structure within loose connective tissue called the terminal ductal units (TDUs), consisting of solid cords of epithelial cells that penetrate into mammary stroma [2]. An accelerated, hormone-dependent expansion of the glandular epithelium occurs at puberty, however the final stages of functional development do not take place until gestation. At the time of pregnancy mammary growth becomes exponential, and driven by pregnancy hormones, giving rise not only to more extensive ductal branching, but also to the development of alveolar structures required for milk production. Mammary alveoli are build by functionally differentiated secretory epithelial cells, showing the ability to synthesize and secrete milk components during lactation. These characteristic structures with hollow cavity are formed in the last stage of mammary gland morphogenesis, termed alveologenesis (Figure 1). Weaning terminates the lactation period causing programmed cell death of a substantial part of the secretory epithelium, which leads to mammary gland involution. The cycle of proliferation, differentiation, and regression can be repeated many times in female's life. That is why mammary gland has become a very good and convenient model for studying processes involved in development and differentiation.

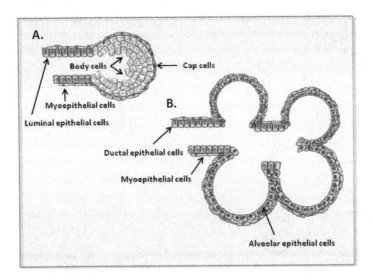

Figure 1. Schematic representation of the structure of mammary epithelium on different stages of differentiation. (A) Prepubertal and pubertal mammary gland consists of ducts terminated by highly proliferative terminal end buds (TEBs) that comprise cap cells in direct contact with the basal lamina, and body cells forming multicellular bulk of the TEB. (B) During pregnancy more extensive ductal branching and formation of alveolar structures required for milk production takes place. In functionally active mammary gland alveoli are built by a single layer of milk secreting luminal epithelial cells surrounded by myoepithelial cells, and basement membrane. Myoepithelial cells contractions release the milk to the ducts, and further to the nipple, whereas basement membrane provides cell contact with extracellular environment.

The lactation cycle includes periods of intensive proliferation of mammary epithelial cells (MECs), their functional differentiation during lactogenesis, and tissue involution caused by death of the secretory cells. Nowadays it is well established, that apoptosis plays a crucial role in all stages of mammary gland development. It is involved in lumen formation during ductal and alveolar morphogenesis, in replacement of cells during lactation, when MECs show high secretory activity, and in the involution of mammary gland. However, proper growth, development and remodelling require also well controlled balance between protein synthesis and organelle biogenesis versus protein degradation and organelle turnover in the cells. Since autophagy is the major cellular pathway for degradation of long-lived proteins and control of the cytoplasmic organelles, this process is particularly important during development and under certain stress conditions.

2. Role of autophagy in mammary gland development

2.1. Role of autophagy in mammogenesis and its relation with apoptosis

During all stages of mammary gland development lumen formation is essential for building functional network of ducts, terminated by milk-producing alveoli at the time of lactation. Lumen formation follows the process of branching morphogenesis, and it is said to be based mainly on the clearance of cells via apoptosis of an inner cell population within newly branched epithelial cords or newly formed acini, creating a cavity [3]. When TEBs invade the mammary stroma they are built by multilayers of epithelial cells, however, the primary ducts behind them posses only a single outer layer of myoepithelial cells and an inner layer of luminal cells surrounding an empty hollow lumen. It was noted that the body cells of TEBs show high rates of apoptosis indicating that this type of programmed cell death contributes to lumen formation [4]. The hypothesis was additionally confirmed with the use of transgenic mice overexpressing an antipoptotic protein Bcl-2 in the mammary gland, because the mammary glands of these rodents showed delayed lumen formation [4].

2.1.1. Autophagy, apoptosis and lumen formation

More information on the mechanisms regulating the lumenization process were obtained in the studies using a three dimensional (3D) cell culture model. This *in vitro* culture system was developed on the basis of the first observations showing, that epithelial cells are able to maintain their tissue structure when grown on ECM components which more closely mimic the *in vivo* microenvironment then the rigid plastic surfaces used in the classic monolayer cultures [5, 6]. It has been shown that mammary epithelial cells (MECs) cultured on laminin-rich reconstituted basement membrane (rBM) are able to recapitulate numerous features of mammary epithelium *in vivo*, including the formation of acini-like spheroids, with a hollow lumen, apicobasal polarization of cells, basal deposition of basement membrane components (collagen IV, laminin I), and the ability to produce milk proteins [5, 7, 8]. Upon seeding within the rBM, normal MECs first undergo a few cycles of proliferation forming small organoids. Next, the structures develop an axis of apicobasal polarity, illustrated among others by basal localization of integrin receptors which are in direct contact with ECM, and lateral (e.g. E-cadherin) or apico-lateral (e.g. ZO-1) localization of junctional complexes [7, 8]. The spherical structures subsequently become unresponsive to proliferative signals, and a bona fide lumen is formed by cavitation, involving the removal of centrally localized cells by death processes [9, 10, 1] (Figure 2). Lack of cell contact with ECM is regarded to be the direct cause of apoptosis initiation in the MECs placed in the centre of developing spheroid. This type of apoptotic death program is termed anoikis [11, 12]. However, it has been shown that overexpression of antiapoptotic proteins Bcl-2 or Bcl-XL in acini formed by human mammary epithelial cell line

MCF-10A only delayed lumen clearance for a few days, although apoptosis was inhibited, which pointed at a possibility that other processes may contribute to lumen formation [9]. In the same study electron microscopy analysis revealed the presence of numerous autophagic vacuoles in the central cells of developing acinar structures. Since autophagy was observed also in spheroids overexpressing Bcl-2 it was concluded that this process proceeds independently of apoptosis. Thus, it was initially proposed that autophagy may also promote lumen formation by initiating type 2 of programmed cell death [9, 13]. However, this hypothesis had some flaws, as it ignored the fact, that the cells lacking contact with ECM may initially induce autophagy as a cytoprotective mechanism against this stressful condition.

2.1.2. Role of autophagy in cells lacking contact with ECM

Integrin receptors are responsible for sustaining cell-matrix interactions, and mediating signal transduction from ECM into the cells [14]. β1-integrin has been shown critical for the alveolar morphogenesis of a glandular epithelium and for maintenance of its differentiated function [15]. Inhibition of β1-integrin in human MECs using blocking antibodies resulted in induction of autophagy. On the contrary, when laminin-rich basement membrane was added to the cells cultured in suspension autophagy was not induced, which points at a direct relationship between autophagy induction and the loss of cell contact with ECM [16]. Furthermore, it was demonstrated that depletion of some of the major Atg genes responsible for autophagy induction and autophagosome formation, namely: Atg5, Atg6 (beclin1) and Atg7, using siRNA technique resulted in reduced autophagy and enhanced apoptosis in suspended cells. A reduced clonogenic viability upon reattachment of MECs was also observed, indicating the prosurvival function of autophagy in cells lacking the direct contact with ECM during acini formation [16]. Interestingly, these studies have shown that inhibition of autophagy either pharmacologically (using 3-MA) or by knocking down Atg5 or Atg7 genes failed to elicit long-term luminal filling even when combined with inhibition of apoptosis. Similar results were obtained by Karantza-Wadsworth and co-workers [17], who worked on immortalized mouse mammary epithelial cells lacking one allel of beclin1 (beclin1$^{+/-}$). These cells when grown in 3D culture formed acini, which exhibited accelerated lumen formation compared to wild type controls that had both alleles of beclin1 gene (beclin1 $^{+/+}$). The authors concluded that defective autophagy may sensitize MECs to metabolic stress, leading to accelerated lumen formation. Central acinar cells of Bcl-2-expressing beclin1$^{+/-}$ spheroids exhibited signs of necrotic cells death, suggesting that necrosis may be the default cell death mechanism upon apoptosis and autophagy inactivation [17]. However, their study also indicated, that defective autophagy compromises the ability of cells to adapt to metabolic stress, which may lead to insufficient ATP generation, accumulation of damaged mitochondria with excessive reactive oxygen species (ROS), and this in turn may cause accumulation of DNA damage, resulting in genome instability and increased risk of cancer progression.

The role of autophagy as the first line survival mechanism of cells centrally localized in the acinar structures was also proven by Sobolewska et al. [18] in the studies on bovine MECs. Bovine BME-UV1 mammary epithelial cells cultured on rBM behave in a similar manner to other described MECs, forming acinar structures composed of an outer-layer of polarized cells and a hollow lumen in the centre of the spheroids within 16 days of 3D culture. Autophagy was observed on the basis of the punctuated pattern of GFP-LC3 protein in the centre of developing acini by the end of the first week of 3D culture. The induction of autophagy preceded apoptosis, as the expression of apoptosis executor enzyme - cleaved caspase-3 was detected starting from the 9th day of cell culture. Thus, autophagy was observed in the acinar structures when a clear distinction of two populations of cells within the structures could be determined – the outer polarized layer of cells with direct contact with rBM, and the centrally localized cells lacking this contact. Subsequent intensive apoptosis eliminated the inner cells forming hollow lumen of the acini [18]. The importance of autophagy and the time of its activation during formation of spherical structures was further confirmed in the experiments on bovine MECs cultured in 3D system in the presence of 3-MA – the inhibitor of early autophagosome formation. 3MA caused formation of small, underdeveloped organoids, and the cells forming these structures showed signs of apoptosis (cleaved caspase-3 activity) before the process of polarization was completed [18]. However, others have shown that the addition of 3-MA to the 3D culture in the later time points, when minimal luminal apoptosis was already observed, caused only increased luminal cell death, not influencing the shape of the acini [16]. Thus, not only localization of autophagic cells, but also the time of autophagy induction, determines the proper development of mammary alveoli

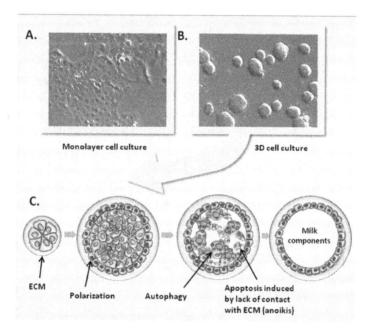

Figure 2. Formation of alveoli-like structures by mammary epithelial cells (MECs). (A) Image of bovine BME-UV1 MECs forming *in vitro* a monolayer on plastic surface. (B) Image of BME-UV1 cells forming 3D spherical structures *in vitro* when cultured on extracellular matrix (ECM) components. (C) Schematic representation of processes that take place

after seeding MECs within ECM: at the beginning of 3D culture cells undergo a few cycles of proliferation, forming small organoids. Next the outer layer of cells in direct contact with ECM develops an axis of apicobasal polarity, while the centrally localized cells lacking the necessary signals from the matrix undergo metabolic changes. At first the inner cells induce autophagy as a survival mechanism, but the sustained stress conditions subsequently lead to lumen formation by apoptotic cell death. The developed alveoli-like structures are able to secrete milk components into the luminal space.

2.1.3. Summary

In the process of mammary alveoli formation the outer layer of epithelial cells, which is in direct contact with ECM undergoes proper apicobasal polarization, and in the later stages develops specific secretory abilities. During alveologenesis autophagy is induced in the cells localized in the centre of the developing alveoli, as a results of the lack of contact of those inner cells with ECM. Autophagy is activated as a survival mechanism under the stress conditions connected with insufficient nutrient and energy supplies, and its main role is cells protection from potential damage of mitochondria and genome instability. The sustained stress conditions in the centre of the alveolar structures lead to apoptosis induction, and elimination of the inner cells by programmed cell death, which results in formation of hollow lumens of the alveoli (Figure 2).

2.2. Extracellular and intracellular factors regulating autophagy in mammary epithelial cells during mammogenesis

2.2.1. Role of endoplasmic reticulum kinase — PERK in the induction of autophagy during alveoli formation

Lack of cell contact with ECM in the centre of developing acinar structures leads to a rapid decrease in glucose intake, which correlates with a drop in ATP levels, and progressive accumulation of reactive oxygen species [19]. In this context the subsequent induction of autophagy supports the hypothesis about the primary adaptive and survival function of this process in the inner population of MECs. Studies on the potential mechanisms taking part in the induction of autophagy during acini formation pointed at the role of endoplasmic reticulum kinase – PERK in this process. PERK kinase is known to attenuate the initiation of translation by phosphorylating eIF2α (eukaryotic initiation factor 2α), when an accumulation of misfolded proteins in endoplasmic reticulum (ER) lumen occurs. It has been shown that upon loss of adhesion MECs activate the canonical PERK-eIF2 α signalling pathway, which serves as an important transcriptional regulator of multiple autophagic genes (ATGs), such as: Atg5, beclin1, Atg8/LC3, involved in autophagosomes formation [20]. PERK not only takes part in the induction of autophagy, but also contributes to the maintenance of ATP production and stimulation of a ROS detoxification response. All together these mechanism protect cells, until the adhesion can be restored, however if the stressful conditions persist the cells finally undergo apoptotic or autophagic death. An evidence for this hypothesis was obtained in the experiment on human MCF10A mammary epithelial cells cultured on rBM. Avivar-Valderas and co-workers [20] observed that enforced PERK activation during the late stages of acinar structures development allowed the centrally localized cells to persistently occupy the luminal space. At the same time an increase in the number of basal cells was noted, suggesting that some of the surviving MECs reattached to ECM in the outer/basal layer of acinar cells. Complementary observations were made *ex vivo* on murine mammary glands isolated at the lactation period. Immunohistochemical analysis of the expression of activated/phosphorylated PERK (p-PERK) and autophagic marker: LC3 in mammary tissue from lactating mice revealed that PERK was highly activated in the cells found in the luminal space of the mammary alveoli, as well as in the luminal epithelium, whereas LC3 was detected only in the detached cells. On the other hand, the expression of pro-apoptotic protein BimEL was weakly detected in the cells found in the luminal space of the mammary tissue, however and increased staining was observed in the epithelium of female mice with conditional deletion of PERK gene. Simultaneously, autophagy was decreased in the tissue samples from the genetically modified animals, suggesting that activation of PERK promotes autophagy and inhibits induction of apoptosis enabling a sustained survival of mammary epithelial cells during lactation [20].

2.2.2. Regulation of autophagy by signalling pathway mediated by mTOR kinase

Another kinase that may be involved in the induction of autophagy in the inner cells of the developing alveoli is AMP-activated protein kinase (AMPK). This enzyme is activated through the upstream kinase LKB1 when the cellular energy levels are reduced due to intracellular metabolic stress, leading to an increased AMP to ATP ratio. Activation of AMPK in turn leads to phosphorylation of the tuberous sclerosis complex (TSC1/2 complex), causing inhibition of mTOR. mTOR (mammalian target of rapamycin) is a conserved Ser/Thr protein kinase that regulates cell growth, cell cycle progression, protein synthesis and nutrient import [21]. In the nutrient and energy rich conditions it is also considered an inhibitor of autophagy [22]. Thus, the reduction of energy levels leads to mTOR inhibition via the LKB1 and AMPK kinases activation, and stimulation of autophagy in the cells [23]. Since AMPK activity was shown to be significantly increased in the MECs lacking contact with ECM it is highly probable that this pathway is also involved in autophagy regulation during alveolar lumen formation.

The activity status of mTOR constitutes an important switch in cell metabolism and fate. As mentioned above, when the energy and nutrient supply is sufficient the signalling pathway activating mTOR and its downstream targets is involved in translation regulation, mRNA turnover, protein stability, actin cytoskeletal organization, cell cycle progression and inhibition of autophagy [21]. This kinase is a target of a macrolide antibiotic called rapamycin, which specifically inhibits mTOR. Rapamycin was used in the studies on the mechanisms regulating the development of acinar structures formed by bovine mammary epithelial cells cultured on rBM [18]. Addition of the drug from the first day of 3D culture resulted in formation of small, underdeveloped spheroids, because rapamycin blocked cell proliferation. At the same time autophagy was induced in all cells forming the acini, as judged by high expression of the active form of LC3-II protein. The induction of autophagy prevented cells from immediate cell death, since the levels of the apotosis executor enzyme – cleaved caspase 3 were reduced in the rapamycin treated acinar structures. Results of this experiment further confirmed the protective role of autophagy in MECs, but also showed the importance of the proper timing of autophagy induction, which should not precede the period of intensive growth of the developing acini and the proper polarization of the outer layer of epithelial cells.

2.2.3. Mitogenic function of IGF-I and its effect on the rate of differentiation of mammary alveolar structures

During the time of mammary gland development several growth factors synthesized locally in the stromal or mesenchymal compartment of the gland, such as: IGF-I (insulin-like growth factor-I), EGF (epidermal growth factor), or HGF (hepatocytes growth factor), induce cell proliferation and survival, leading to expansion of the glandular epithelium. IGF-I plays a pivotal role in mammary tissue homeostasis, stimulating cell proliferation and differentiation during mammogenesis and lactogenesis. This growth factor exerts both endocrine and local actions. It is produced in the liver in response to pituitary growth hormone (GH), but is also synthesized and secreted by the cells of the mammary gland. Signals from IGF-I are transmitted into the cell via type I IGF receptor (IGFIR) located on the surface membranes of epithelial cells. When IGF-I binds to its receptor, IGFIR associates with the p85α subunit of phosphatidylinoinositol-3-kinase (PI3K) and activates another downstream target – Akt kinase [24]. Active Akt (phosphorylated at Ser^{473}) initiates other downstream signalling components involved in initiation of proliferation or activation of survival mechanisms. Moreover, the effect of IGF-I on cell growth and metabolism, mediated by Akt involves also activation of mTOR signalling pathway. Therefore, during normal mammary gland development the mitogenic signals from IGF-I must be under control of other locally produced growth factors and systemic hormones in order to maintain the proper homeostasis in the mammary gland. In fact, studies have shown that mammary epithelial cells grown on ECM components in the presence of IGF-I formed large spheroids lacking a hollow lumen in the centre [18]. The MECs showed prolonged proliferative activity, and decreased apoptosis measured on the basis of cleaved caspase-3 expression. At the same time an increased autophagy was observed in the centrally localized cells. The intensive autophagy of these inner cells, however, might have been induced by stressful conditions evoked by the lack of contact with ECM components, and decreased availability of nutrients inside of the large organoids, rather than directly by IGF-I. In fact, another study with the use of human mammary epithelial cell line MCF-10A over-expressing IGFIR, showed that the cells formed large, misshapen acinar structures with filled lumens and disrupted apico-basal polarisation in the presence of IGF-I [25]. The investigators observed that the MECs over-expressing IGFIR showed increased proliferation and decreased apoptosis, which was connected with increased activity of Akt, as well as mTOR. The phenotype of large misshapen spheroids could also be obtained, when MCF10A cells expressed a conditionally active variant of Akt [26]. Sustained Akt activation caused enhanced proliferation, and increased cell size, along with variability in size and shape of the cells forming the large spheroids. However, when rapamycin was added to the 3D culture the morphological disruption was prevented, indicating that mTOR function is required for the biological effect of Akt action during acinar development, and that the activity of mTOR also needs to be tightly regulated [26]. Although the described studies did not examine the role of autophagy in these processes one can expect that the effect of rapamycin addition not only resulted in inhibition of the proliferative signals induced by conditionally active Akt, but also induced autophagy, which could participate in lumen clearance.

2.2.4. 17β-estradiol and progesterone control mammogenesis by stimulation of mammary epithelial cells proliferation, regulation of gene expression, and induction of autophagy

The proper development of mammary gland is possible thanks to interactions of many cellular signalling pathways induced by intramammary factors, as well as endocrine hormones. Sex steroids belong to the important regulators of normal mammogenesis. Throughout puberty and gestation 17β-estradiol (E2) and progesterone (P4) induce proliferation of mammary epithelium and act as survival factors. Biological responses induced by both hormones are mediated by their receptors, which are located inside the cells, and translocate from the cytoplasm to the nucleus upon activation. There are two types of estrogen receptors (ERα, and ERβ), however ERα is shown to play the major function in the mammary gland. The expression of ERα was found both in the epithelial and stromal compartment of the mammary gland in many species, although in humans and heifers only mammary epithelium express ER. Progesterone can also act through two types of specific receptors: PR-A and PR-B, and the ratio in expression of both isoforms in the mammary gland is critical for the normal response to P4 [27]. PR is expressed only by mammary epithelial cells, and not all MECs show its expression, thus a paracrine interaction occurs between the PR-positive and PR-negative cells, which activates the proliferative and survival signals.

E2 is considered to be responsible for ductal morphogenesis, while P4 is critical for lobulo-alveolar development and transition from ductal to lobulo-alveolar morphology [27]. However, E2 also plays an indirect role in the alveologenesis by stimulating the PR expression in mammary epithelial cells [28]. Additionally, recent studies using the 3D culture system have shown that sex steroids may be involved in autophagy induction during mammogenesis [18]. When bovine MECs were cultured on rBM in the presence of E2 or P4 the cells formed proper acinar structures. During the development of these acini an intensified induction of autophagy was observed in the centre of the structures, as judged by the higher fluorescence intensity of the condensed pattern of GFP-LC3 autophagy marker. Additionally, apoptosis was also elevated in these cells, which led to a faster formation of the hollow lumen inside the spherical structures. Moreover, in case of 17β-estradiol it was shown that this hormone not only accelerated formation of the membrane-bound form of LC3 (LC3-II), but also increased the level of the LC3-I protein [18]. It is well established that both sex steroids exert an influence on target cells through a genomic pathway after binding to their receptors that translocate to the nucleus. Inside the nucleus the activated receptors associate with co-activators, or co-repressors, and finally regulate gene transcription by binding to target genes on the specific sites of the promoter regions, called response elements (ERE – estrogen response element, PRE – progesterone response element) [29, 30]. The observed increase in the total amount of LC3 protein indicates that E2 could enhance the expression of *LC3* gene. More recently steroids, especially estrogen (E2), have been found to exert rapid, non-genomic effects via membrane-bound receptors (mER), causing stimulation of cytoplasmic signalling pathways, such as: MAPK, and PI3K/Akt [31, 30]. So far the non-genomic molecular mechanism of steroids has not been investigated in regard to their possible influence on autophagy induction. However, John and co-workers [32] reported that beclin1 is able to bind with ERα, and the interactions between these proteins may modulate their action. Thus, sex steroid play a major role in the control of mammary gland development not only by acting as prosurvival factors, and stimulating epithelial cells proliferation, but also by regulating autophagy during alveoli formation. Furthermore, it is possible that E2 and P4 may regulate the action of other intra-mammary factors in the mammary epithelium by interactions of the signalling pathways induced by these endocrine and local factors (Figure 3).

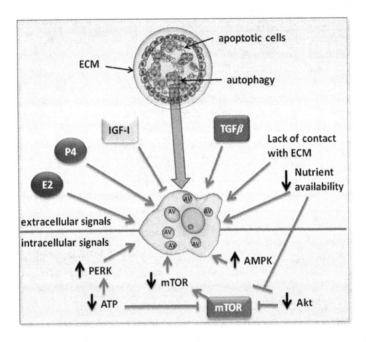

Figure 3. Extracellular and intracellular factors regulating autophagy induction in mammary epithelial cells during the process of lumen formation in mammary alveoli.

2.2.5. Summary

Induction of autophagy in the centrally localized cells of developing acini is regulated by several intracellular pathways and extracellular factors. The stress caused by insufficient nutrient and energy supply in the centre of the alveolar structures activates PERK kinase, which is an important transcriptional regulator, controlling the expression of autophagic (Atg) genes. Increased intracellular metabolic stress inside the developing alveoli also activates AMPK kinase, which inhibits the mTOR mediated signalling pathway leading to autophagy induction. During mammogenesis, the balance between proliferation and cell death processes is also controlled by locally secreted growth factors (i.a. IGF-I), and endocrine hormones (i.a.

sex steroids). IGF-I is an important survival factor, involved in stimulation of the enhanced growth of mammary epithelium during mammogenesis, whereas E2 and P4 play a major role in the control of mammary gland development by stimulating epithelial cells proliferation, as well as regulating autophagy induction during alveoli formation (Figure 3).

2.3. Role of autophagy in mammary gland involution

Mammary gland shows full functional activity during lactation, when the lobules contain fully developed alveoli formed by differentiated mammary epithelial cells secreting milk components into the luminal space. The surrounding myoepithelial cells contract, releasing the milk further to the ductal network, which delivers the milk to the nipple. The process of milk synthesis is under control of galactopoetic hormones (prolactin – PRL, growth hormone – GH), which stimulate the expression of milk proteins, survival of the glandular epithelium and contractions of the alveoli (oxitocin). After the period of functional activity, when females stop feeding their offspring the mammary gland regresses and returns to the state of development similar to the one prior pregnancy. This stage of remodelling is termed involution. Involution can be gradually initiated in the mammary gland, starting from the peak of lactation, because the young are progressively weaned. In case of dairy animals it starts with the natural, progressive decline in the milk yield. Alternatively, the mammary gland involution can be induced by litter removal (forced weaning), or in diary animals by termination of milking.

The withdrawal of suckling or cessation of milking results in the interruption of the release of galactopoetic hormones, which leads to milk stasis, and a rapid decline in milk synthesis caused by downregulation of genes involved in this process. In rodents, which have been extensively used as a model for studying the progress of mammary gland involution, forced weaning very quickly (within 24h after pup removal) leads to the first signs of apoptotic cell death of the epithelium, as some of the MECs are shed into the lumens of the alveoli. This stage, however, is still reversible, and the renewal of suckling preserves the structure of the secretory tissue. When the involution progresses several other processes take place, leading to the regression of the glandular epithelium.

At the time when the first apoptotic cells can be observed in the luminal space, MECs which remain within the alveoli begin reabsorbing the residual milk. Additionally, it has been shown that these cells undergo a change in their phenotype from secretory to phagocytic, which enables them to actively reabsorb also the apoptotic cells from the lumens by a process resembling efferocytosis [33]. During efferocytosis the cell membrane of phagocytic cell engulfs the apoptotic cell, forming a large fluid-filled vesicle, called efferosome or phagosome, which contain the dead cell. The efferosome subsequently fuses with lysosome, causing degradation of the engulfed material. The change in the phenotype of MECs requires changes in the expression of many genes, and thus, is thought to be transcriptionally-mediated. It has been shown, that more than 20 traditional markers of lysosomal activity are upregulated within 24h of forced weaning, and LC3 was detected among these upregulated proteins [33]. Since it is one of the key proteins involved in autophagosomes formation and fusion of lysosomes with autophagic vacuoles, these results indicate that autophagy is induced during the early stages of involution. Although there is no additional information on the role of autophagy during the initial phase of involution, there are evidence showing participation of this process when the regression of the mammary gland progresses.

In the second phase of involution proteolysis of the extracellular matrix and further apoptosis of the secretory epithelium takes place, causing the alveoli to collapse. There is an increase in expression of the protease genes, such as plasminogen activators (serine proteases), that induce the formation of active plasmin from plasminogen. Subsequently, plasmin activates matrix metalloproteinases (MMPs), which are responsible for the proteolytic degradation of basement membrane and ECM of the mammary gland. Removal of ECM induces apoptosis of the epithelial cells, that failed to respond to the first phase of apoptotic signals. The large number of apoptotic cells and debris are removed by phagocytosis performed by professional and non-professional phagocytes (macrophages, and epithelial cells, respectively) [34]. Finally, in the last stage of involution the regrowth of stromal adipose tissue is observed, filling the space of the regressed epithelium. The described course of mammary gland regression concerns the situation when lactation is separated from gestation by a dry period, during which the mammary gland remains in a quiescent state. It was extensively studied in rodents, and is often considered to reflect the general changes during the remodelling of the mammary gland in mammalian species. However, it is well documented that the mammary gland involution in ruminants differs in a significant manner. In cows and goats there is a characteristic overlap between the periods of lactation and next pregnancy, which means that these animals are typically pregnant when the involution is induced by termination of milking. Thus, the high levels of pregnancy hormones stimulating the development of new secretory tissue oppose the stimuli for mammary involution initiated by the milk stasis.

2.3.1. Role of autophagy in the regenerative involution of ruminant mammary gland

The nonlactating period before parturition in diary animals is termed the dry period. During this time the morphological changes in the mammary gland of ruminants (especially cows) are less pronounced than those occurring in the involuting glands of mice or rats. They reflect the change in the secretory state of the mammary epithelium, rather than the characteristic features of the tissue regression. The alveolar structure remains mostly intact during bovine mammary gland involution, even after several weeks of the dry period, although about 30 days after milking cessation the luminal area in the mammary tissue decrease and epithelial cells exhibit few secretory vacuoles [35]. Some of the bovine MECs undergo apoptotic cell death during involution, however, this subpopulation is significantly smaller than in rodents. It is considered that the nonlactating period in dairy cows serves to enhance the replacement of the senescent mammary epithelial cells prior to the next lactation, and thus, the processes taking place during that time are described as regenerative involution. Interestingly, studies have shown that bovine mammary tissue during the dry period shows signs of autophagy. It is manifested by increased expression of beclin1 and a high number of cells with typical morphological features of autophagy (autophagosomes and autophagolysosomes) [36]. Furthermore, *in vitro* studies on BME-UV1 bovine mammary epithelial cells revealed that cells partially devoid of nutritional factors and bioactive compounds induce formation of the autophagosome membrane-bound LC3-II form [37]. These experiments aimed at reflecting the conditions observed in the bovine mammary gland during dry period, when enhanced competition of intensively developing fetus and mother organism for nutritional and bioactive compounds creates a state of temporary malnutrition of MECs. When the concentration of fetal bovine serum (FBS) was significantly reduced in the culture medium of BME-UV1 cells (from standard 10% to 0.5%), the activity of mTOR kinase was significantly decreased, which corresponded with the induction of autophagy. Moreover, autophagy induced by FBS-withdrawal was inhibited by an addition of IGF-I, or EGF. Both growth factors play a prosur-vival role in MECs during mammary development, whereas at the dry period their activity is decreased, similarly to the decreased levels of lactogenic hormones. Simultaneously the levels of sex steroids are elevated due to the pregnancy overlapping the period of glandular involu-tion. The *in vitro* studies demonstrated, that in the presence of E2 or P4 bovine MECs cultured in FBS-deficient conditions showed higher levels of autophagy, which suggests, that these hormones additionally stimulate the induction of this process [37]. Thus, autophagy may be induced in bovine mammary epithelial cells as an additional survival process, which partici-pates in preservation of the glandular morphology during involution.

Additionally, it was shown that autophagy can also be stimulated by transforming growth factor – beta 1 (TGF-β1), a cytokine classified as local growth inhibitor and apoptosis inducer in many cell types, including MECs. TGF-β1 expression was shown to be high during puberty and involution, low during gestation, and undetectable at the time of lactation. This growth factor can regulate cellular processes by a specific signalling pathway, which is induced upon binding of TGF-β1 to its membrane receptors (TβRI and TβRII). The receptors then form heterocomplexes and activate downstream components - the Smad proteins [38]. Smads transmit the signal to the nucleus, where they play a role of transcription factors and bind to DNA on the promoter region regulating the transcription of specific genes. TGF-β1 was shown to regulate the expression patterns of cyclins involved in the cell cycle progression, cell adhesion elements, such as integrins, and IGF binding proteins (IGFBP-3,4 and 5), which regulate the activity of IGF within the mammary gland [27, 39]. For example, IGFBP5 prevents binding of IGF-I to its receptor and inhibits the prosurvival signals. TGF-β1 is also able to induce apoptosis in mammary epithelial cells through the mitochondrial pathway involving: activation and translocation of the proapoptotic protein Bax to mitochondrial membranes, release of cytochrome c, and activation of the executor enzyme caspase-3 [40]. The experiments on bovine MECs revealed, that this cytokine also increased the level of LC3 and beclin1 proteins, indicating the direct role of TGF-β1 in autophagy induction. Moreover, it was found that the high expression of TGF-β1 receptors in the involuting bovine mammary tissue correlated with increased levels of beclin1 and downregulation of growth hormone receptor (GH-R) and IGF-I receptor (IGF-IRα) in this tissue [36].

The induction of autophagy by TGF-β1 was also observed during mammary acini formation in the studies with the use of 3D culture system [18]. These results correspond with other findings, showing that TGF-β1 is responsible for regulation of growth and pattering of the mammary ductal tree during mammogenesis, and can partially act by modulation of the effect of IGF-I on the developing tissue [41, 39].

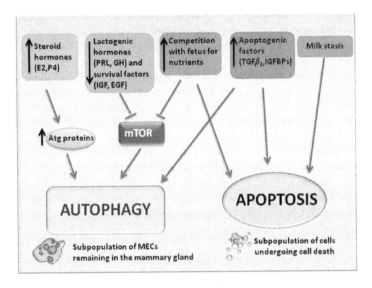

Figure 4. Endocrine hormones, auto/paracrine factors and intramammary conditions inducing autophagy and apoptosis in bovine mammary epithelial cells during regenerative involution.

2.3.2. Summary

In the period of mammary gland involution, during which regression of the secretory epithelium takes place returning the gland to the quiescent state, autophagy seems to be involved in the efferocytosis of the apoptotic epithelial cells. In case of the regenerative involution of bovine mammary gland autophagy is induced as a survival mechanism participating in the preservation of the glandular epithelium prior to next lactation. When milking is terminated a small population of mammary epithelial cells undergo apoptotic cell death, due to milk stasis and increased levels of proapoptotic factors, such as: TGF-β1. The remaining cells down-regulate milk secreting pathways and await parturition by inducing autophagy, as a mechanism which stabilizes intracellular supplies of energy and amino acids at the time of enhanced competition of intensively developing fetus and mother organism for nutritional and bioactive compounds. Additionally, local factors, such as TGF-β1, and pregnancy hormones (17β-estradiol and progesterone) stimulate autophagy during the dry period, suggesting a possible role of these factors in the control of the balance between apoptosis and survival of the epithelial cells in the involuting bovine mammary gland (Figure 4).

3. Autophagy and breast cancer

Most of breast malignancies arise in the terminal duct lobular units (TDLUs). In general, carcinomas are characterized by the loss of epithelial polarity and tissue organization. Cancer cells, which remain within the basement membrane of the mammary ductal-lobular system are classified as benign in situ carcinomas, whereas when neoplatic cells invade into the adjacent stroma the tumour becomes malignant [42]. Early premalignant breast cancer lesions, such as hyperplastic lesions with atypia and carcinoma *in situ* are characterized by a complete or partially filled lumen [1]. Moreover, *in vitro* experiments with the use of 3D culture system have shown, that human breast tumour cell lines are not able to form acinar structures with the centrally localized hollow lumen, and polarisation of cells surrounding this lumen. Instead they develop into nonpolarized clusters with limited differentiation [43, 44]. Both apoptosis and autophagy have been shown to be involved in the process of lumen clearance, however, autophagy is thought to be induced first as a survival mechanism in the central acinar cells, which are under increased metabolic stress connected with the lack of contact with ECM, hypoxia and decreased nutrient and energy supplies. Karantza-Wadsworth and co-workers [17] have shown that monoallelic deletion of beclin1 (beclin1$^{+/-}$) that leads to defective autophagy, causes increased DNA damage, and genome instability in cells. When defective autophagy is synergized with defects in apoptosis machinery, mammary tumorigenesis can be promoted.

Perturbations in autophagy has been implicated in the pathogenesis of diverse disease states, including cancer. The monoallelic deletion of beclin1 is observed in 50% of breast tumours [45]. Human breast carcinoma cell lines, as well as tumour tissue have decreased beclin1 levels, while mammary tissue from beclin1$^{+/-}$ mice shows hyperproliferative, preneoplastic changes [46]. Moreover, beclin1$^{+/-}$ immortalized mouse mammary epithelial cells, which exhibit compromised autophagy under metabolic stress, cause accelerated tumorigenesis after allogeneic transplantation into nude mice [17]. On the contrary, beclin1 ectopic expression in MCF-7 breast cancer cells, which are tetraploid but have only three beclin1 copies, led to a slower proliferation of these cells *in vitro*, as well as *in vivo* in the xenograft tumours. These findings indicate that beclin1 is a haploinsufficient tumour suppressor [47]. Furthermore, when deficiency in autophagy synergizes with defective apoptosis the response to the environmental stress is impaired and tumorigenecity is increased, promoting tumour growth.

On the other hand, autophagy as a known survival mechanism preserves cell viability during periods of nutrient limitation and hypoxia, which suggests that it can sustain cellular metabolism within the tumour. Metabolic stress is a common feature of solid tumours, resulting from inadequate vascularisation, and causing nutrient, growth factors, and oxygen deprivation [48]. It has been shown that solid tumours formed by cells with defective apoptosis are able to survive the metabolic stress by inducing autophagy [49]. Thus, when tumour cells have intact autophagy it may be induced as an adaptive response to anticancer agents, in which case autophgay may act as a treatment resistance mechanism prolonging tumour cell survival. It is especially important in apoptosis-defective cancers, which rely on autophagy under stressful conditions. In this case inhibition of autophagy should inhibit cancer cells' survival and enhance the efficacy of anticancer treatment [45]. Attempts to use autophagy inhibitors to sensitize cancer cells to treatment have been recently reported. For example, knockdown of autophagic genes in MCF-7 and T-47D breast cancer cells, combined with tamoxifen or 4-hydroxy-tamoxifen treatment, resulted in decreased viability of these cells [50, 51]. Hydroxychloroquine, which is a lysosomotropic agent causing increase in intralysosomal pH, and impairing autophagic protein degradation, has been used in clinical trials to modulate autophagy in metastatic breast cancer [45]. On the contrary, autophagy-deficient cancer cells with intact apoptotic machinery are shown to be particularly sensitive to therapeutically agents inducing metabolic stress (e.g.anti-angiogenic drugs), as these drugs cause apoptosis of the tumour cells. However, in a situation, when tumour cells show defective autophagy and apoptosis the approach to treatment should be different. As mentioned previously, simultaneous deficiency in autophagy and apoptosis makes the tumour cells susceptible to metabolic stress and DNA damage, leading to genome instability. Thus, use of metabolic or replication stress-inducing agents may cause further DNA damage in these cells, resulting in enhanced tumorigenic potential and development of drug resistance [17].

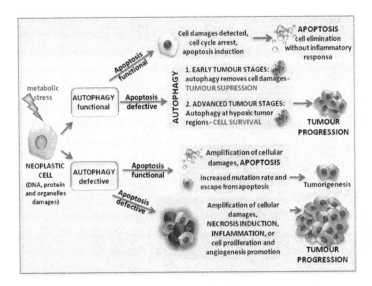

Figure 5. Different scenarios of response to metabolic stress in mammary tumorigenesis, depending on the functional status of autophagy and apoptosis in the tumour cells.

Finally, recent studies have pointed at a possible, yet still unexplored role of autophagy during invasion and metastasis. The studies with the use of 3D culture system revealed, that autophagy is induced in cell lacking the direct contact with ECM. The ability to survive in the absence of normal ECM is critical for metastasis, since cancer cells in the bloodstream or secondary tissue sites are either deprived of matrix or exposed to foreign matrix components [52]. The metastatic, secondary tumours are often resistant to therapy. Disseminated tumour cells, prior to development into the secondary tumours can remain in a dormant state for many years. *In vitro* and *in vivo* studies on mice have shown, that inhibition of β1-integrin, one of the subunits of integrin receptor responsible for cell contact with ECM, prevents proliferation of tumour cells, but not their viability, leading to induction of dormant state [53, 54]. Other studies have shown that cell detachment from ECM induces autophagy [16, 18], and blocking β1-integrin function in attached human MECs is also sufficient for autophagy induction [16]. Therefore it is possible that detachment-induced autophagy in disseminated tumour cells may promote survival of these cells in the dormant state.

All presented results show the complexity of the role of autophagy in cancer development and progression. The possible role of activated, as well as defective autophagy in tumorigenesis is summarized in Figure 5. Research work continues to determine the molecular pathways regulating autophagy in tumour cells on different stages of tumour development. Results of the future studies may be beneficial for proper modulation of autophagy during cancer treatment and prevention.

Nomenclature

AMPK – AMP-activated protein kinase

E2 - 17β-estradiol

ECM – extracellular matrix

eIF2α – eukaryotic translation initiation factor 2 alpha

ER – endoplasmic reticulum

GFP-LC3 – LC3 (Atg8) protein fused with green fluorescence protein (a marker of autophagy)

GH – growth hormone

IGFBPs - insulin growth factor binding proteins

IGF-I – insulin-like growth factor-I

MECs – mammary epithelial cells

mTOR – mammalian target of rapamycin

P4 – progesterone

PERK – endoplasmic reticulum kinase

rBM – reconstituted basement membrane

TDLU - terminal duct lobular unit

TDU – terminal ductal unit

TEB – terminal end bud

TGF- β1 – transforming growth factor beta1

TSC1/2 – tuberous sclerosis complex (taking part in inhibition of mTOR kinase pathway)

Author details

Malgorzata Gajewska*, Katarzyna Zielniok and Tomasz Motyl

*Address all correspondence to: malgorzata_gajewska@sggw.pl

Department of Physiological Sciences, Faculty of Veterinary Medicine, Warsaw University of Life Sciences, Warsaw, Poland

References

[1] Reginato, M. J, & Muthuswamy, S. K. Illuminating the center: mechanisms regulating lumen formation and maintenance in mammary morphogenesis. Journal of Mammary Gland Biology and Neoplasia (2006). , 11, 205-211.

[2] Capuco, A. V, & Ellis, S. Bovine mammary progenitor cells: current concepts and future directions. Journal of Mammary Gland Biology and Neoplasia (2005).

[3] Mailleux, A. A, Overholtzer, M, & Brugge, J. S. Lumen formation during mammary epithelial morphogenesis: insights from in vitro and in vivo models. Cell Cycle. (2008).

[4] Humphreys, R. C, Krajewska, M, Krnacik, S, Jaeger, R, Weiher, H, Krajewski, S, Reed, J. C, & Rosen, J. M. Apoptosis in the terminal endbud of the murine mammary gland: a mechanism of ductal morphogenesis. Development (1996). , 122(12), 4013-4022.

[5] Barcellos-hoff, M. H, Aggeler, J, Ram, T. G, & Bissell, M. J. Functional differentiation and alveolar morphogenesis of primary mammary cultures on reconstituted basement membrane. Development (1989). , 105, 223-235.

[6] Li, M. L, Aggeler, J, Farson, D. A, Hatier, C, Hassell, J, & Bissell, M. J. Influence of a reconstituted basement membrane and its components on casein gene expression and secretion in mouse mammary epithelial cells. Proceedings of the National Academy of Sciences U S A. (1987).

[7] Debnath, J, Muthuswamy, S. K, & Brugge, J. S. Morphogenesis and oncogenesis of MCF-10A mammary epithelial acini grown in three-dimensional basement membrane cultures. Methods (2003). , 30, 256-268.

[8] Kozlowski, M, Wilczak, J, Motyl, T, & Gajewska, M. Role of extracellular matrix and prolactin in functional differentiation of bovine BME-UV1 mammary epithelial cells. Polish Journal of Veterinary Sciences (2011).

[9] Debnath, J, Mills, K. R, Collins, N. L, Reginato, M. J, Muthuswamy, S. K, & Brugge, J. S. The role of apoptosis in creating and maintaining luminal space within normal and oncogene-expressing mammary acini. Cell (2002)., 111, 29-40.

[10] Shaw KRMWrobel CN, Brugge JS. Use of three-dimensional basement membrane cultures to model oncogene-induced changes in mammary epithelia. Journal of Mammary Gland Biology and Neoplasia (2004)., 9, 297-310.

[11] Frisch, S. M, & Francis, H. Disruption of epithelial cell-matrix interactions induces apoptosis. Journal of Cell Biology (1994).

[12] Gilmore, A. P. Anoikis. Cell Death and Differentiation. (2005). Suppl, 2, 1473-1477.

[13] Mills, K. R, Reginato, M, Debnath, J, Queenan, B, & Brugge, J. S. Tumor necrosis factor related apoptosis-inducing ligand (TRAIL) is required for induction of autophagy during lumen formation in vitro. Proceedings of the National Academy of Sciences U S A. (2004)., 101, 3438-3443.

[14] Hynes, R. O. Integrins: bidirectional, allosteric signaling machines. Cell. (2002)., 110, 673-687.

[15] Naylor, M. J, Li, N, Cheung, J, Lowe, E. T, Lambert, E, Marlow, R, Wang, P, Schatzmann, F, Wintermantel, T, Schüetz, G, Clarke, A. R, Mueller, U, Hynes, N. E, & Streuli, C. H. Ablation of beta1 integrin in mammary epithelium reveals a key role for integrin in glandular morphogenesis and differentiation. Journal of Cell Biology (2005).

[16] Fung ChLock R, Gao S, Salas E, Debnath J. Induction of autophagy during Extracellular Matrix detachment promotes cell survive. Molecular Biology of the Cell (2008)., 19, 797-806.

[17] Karantza-wadsworth, V, Patel, S, Kravchuk, O, Chen, G, Mathew, R, Jin, S, & White, E. Autophagy mitigates metabolic stress and genome damage in mammary tumorigenesis. Genes and Development (2007)., 21, 1621-1635.

[18] Sobolewska, A, Motyl, T, & Gajewska, M. Role and regulation of autophagy in the development of acinar structures formed by bovine BME-UV1 mammary epithelial cells. European Journal of Cell Biology (2011)., 90, 854-864.

[19] Schafer, Z. T, Grassian, A. R, Song, L, Jiang, Z, Gerhart-hines, Z, Irie, H. Y, Gao, S, Puigserver, P, & Brugge, J. S. Antioxidant and oncogene rescue of metabolic defects caused by loss of matrix attachment. Nature. (2009)., 461(7260), 109-113.

[20] Avivar-valderas, A, Salas, E, Bobrovnikova-marjon, E, Diehl, J. A, Nagi, C, Debnath, J, & Aguirre-ghiso, J. A. PERK integrates autophagy and oxidative stress responses to promote survival during extracellular matrix detachment. Molecular and Cellular Biology (2011).

[21] Wullschleger, S, Loewith, R, & Hall, M. N. TOR signaling in growth and metabolism. Cell. (2006).

[22] Pattingre, S, Espert, L, Biard-piechaczyk, M, & Codogno, P. Regulation of macroautophagy by mTOR and Beclin 1 complexes. Biochimie. (2008).

[23] Liang, J, Shao, S. H, Xu, Z. X, Hennessy, B, Ding, Z, Larrea, M, Kondo, S, Dumont, D. J, Gutterman, J. U, Walker, C. L, Slingerland, J. M, & Mills, G. B. The energy sensing LKB1-AMPK pathway regulates kip1) phosphorylation mediating the decision to enter autophagy or apoptosis. Nature Cell Biology (2007). , 27.

[24] Fleming, J. M, Desury, G, Polanco, T. A, & Cohick, W. S. Insulin growth factor-I and epidermal growth factor receptors recruit distinct upstream signaling molecules to enhance AKT activation in mammary epithelial cells. Endocrinology. (2006).

[25] Yanochko, G. M, & Eckhart, W. (2006). Type I insulin-like growth factor receptor over-expression induces proliferation and anti-apoptotic signaling in a three-dimensional culture model of breast epithelial cells. Breast cancer Research 8, R18.

[26] Liu, H, Radisky, D. C, Wang, F, & Bissell, M. J. Polarity and proliferation are controlled by distinct signaling pathways downstream of PI3-kinase in breast epithelial tumor cells. Journal of Cell Biology (2004). , 164, 603-612.

[27] Lamote, I, Meyer, E, Massart-leen, A. M, & Burnvenich, C. Sex steroids and growth factors in the regulation of mammary gland proliferation, differentiation and involution. Steroids (2004). , 69, 145-159.

[28] Atwood, C. S, Hovey, R. C, Glover, J. P, Chepko, G, Ginsburg, E, Robison, W. G, & Vonderhaar, B. K. Progesterone induces side-branching of the ductal epithelium in the mammary glands of peripubertal mice. Journal of Endocrinology (2000).

[29] Nilsson, S, Mäkelä, S, Treuter, E, Tujague, M, Thomsen, J, Andersson, G, Enmark, E, Pettersson, K, Warner, M, & Gustafsson, J. A. Mechanisms of estrogen action. Physiological Reviews (2001). , 81, 1535-1565.

[30] Fox, E. M, Andrade, J, & Shupnik, M. A. Novel actions of estrogen to promote proliferation: integration of cytoplasmic and nuclear pathways. Steroids (2009). , 74, 622-627.

[31] Levin, E. R. Bidirectional signaling between the estrogen receptor and the epidermal growth factor receptor. Molecular Endocrinology (2003). , 17, 309-317.

[32] John, S, Nayvelt, I, Hsu, H. C, Yang, P, Liu, W, Das, G. M, Thomas, T, & Thomas, T. J. Regulation of estrogenic effects by beclin1 in breast cancer cells. Cancer Research (2008). , 68, 7855-7863.

[33] Monks, J, & Henson, P. M. Differentiation of the mammary epithelial cell during involution: implications for breast cancer. Journal of Mammary Gland Biology and Neoplasia (2009).

[34] Monks, J, Geske, F. J, Lehman, L, & Fadok, V. A. Do inflammatory cells participate in mammary gland involution? Journal of Mammary Gland Biology and Neoplasia (2002).

[35] Capuco, A. V, & Akers, R. M. Mammary involution in diary Animals. Journal of Mammary Gland and Neoplasia. (1999). , 4(2), 137-144.

[36] Zarzynska, J, Gajkowska, B, Wojewódzka, E, Dymnicki, E, & Motyl, T. Apoptosis and autophagy in involuting bovine mammary gland is accompanied by up-regulation of TGF-β1 and suppression of somatotropic pathway. Polish Journal of Veteterinary Sciences (2007). , 10, 1-9.

[37] Sobolewska, A, Gajewska, M, Zarzynska, J, Gajkowska, B, & Motyl, T. IGF-I, EGF and sex steroids regulate autophagy in bovine mammary epithelial cells via the mTOR pathway. European Journal of Cell Biology (2009). , 88, 117-130.

[38] Mulder, K. M. Role of Ras Mapks in TGF-beta signaling. Cytokine and Growth Factor Reviews (2000). , 11, 23-35.

[39] Gajewska, M, & Motyl, T. IGF-binding proteins mediate TGF-βinduced apoptosis in bovine mammary epithelial BME-UV1 cells. Comperative Biochemistry and Physiology Part C (2004). , 1.

[40] Kolek, O, Gajkowska, B, Godlewski, M. M, & Motyl, T. Colocalization of apoptosis-regulating proteins in mouse mammary epithelial HC11 cells exposed to TGF-β1. European Journal of Cell Biology (2003). , 82, 303-312.

[41] Daniel, C. W, Robinson, S, & Silberstein, G. B. The transforming growth factors beta in development and functional differentiation of the mouse mammary gland. Advances in Experimental Medicine and Biology (2001). , 501, 61-70.

[42] Weigelt, B, & Bissell, M. J. Unraveling the microenvironmental influences on the normal mammary gland and breast cancer. Seminars in Cancer Biology (2008).

[43] Petersen, O. W, Rønnov-jessen, L, Howlett, A. R, & Bissell, M. J. Interaction with basement membrane serves to rapidly distinguish growth and differentiation pattern of normal and malignant human breast epithelial cells. Proceedings of the National Academy of Sciences U S A. (1992).

[44] Debnath, J, & Brugge, J. S. Modelling glandular epithelial cancers in three-dimensional cultures. Nature Reviews Cancer (2005).

[45] Chen, N, & Karantza-wadsworth, V. Role and regulation of autophagy in cancer. Biochimica et Biophysica Acta. (2009).

[46] Qu, X, Yu, J, Bhagat, G, Furuya, N, Hibshoosh, H, Troxel, A, Rosen, J, Eskelinen, E. L, Mizushima, N, Ohsumi, Y, Cattoretti, G, & Levine, B. Promotion of tumorigenesis by heterozygous disruption of the beclin 1 autophagy gene. Journal of Clinical Investigation (2003).

[47] Yue, Z, Jin, S, Yang, C, Levine, A. J, & Heintz, N. Beclin 1, an autophagy gene essential for early embryonic development, is a haploinsufficient tumor suppressor. Proceedings of the National Academy of Sciences U S A. (2003).

[48] Jin, S, & White, E. Role of autophagy in cancer: management of metabolic stress. Autophagy. (2007).

[49] Degenhardt, K, Mathew, R, Beaudoin, B, Bray, K, Anderson, D, Chen, G, Mukherjee, C, Shi, Y, Gélinas, C, Fan, Y, Nelson, D. A, Jin, S, & White, E. Autophagy promotes tumor cell survival and restricts necrosis, inflammation, and tumorigenesis. Cancer Cell. (2006).

[50] Qadir, M. A, Kwok, B, Dragowska, W. H, To, K. H, Le, D, Bally, M. B, & Gorski, S. M. Macroautophagy inhibition sensitizes tamoxifen-resistant breast cancer cells and enhances mitochondrial depolarization. Breast Cancer Research and Treatment (2008).

[51] Samaddar, J. S, Gaddy, V. T, Duplantier, J, Thandavan, S. P, Shah, M, Smith, M. J, Browning, D, Rawson, J, Smith, S. B, Barrett, J. T, & Schoenlein, P. V. A role for macroautophagy in protection against hydroxytamoxifen-induced cell death and the development of antiestrogen resistance. Molecular Cancer Therapeutics (2008). , 4.

[52] Debnath, J. Detachment-induced autophagy during anoikis and lumen formation in epithelial acini. Autophagy (2008). , 4, 352-335.

[53] White, D. E, Kurpios, N. A, Zuo, D, Hassell, J. A, Blaess, S, Mueller, U, & Muller, W. J. Targeted disruption of betaintegrin in a transgenic mouse model of human breast cancer reveals an essential role in mammary tumor induction. Cancer Cell. (2004). , 1.

[54] Lock, R, & Debnath, J. Extracellular matrix regulation of autophagy. Current Opinion in Cell Biology (2008). , 20, 583-588.

Chapter 8

Role of Autophagy in the Ovary Cell Death in Mammals

M.L. Escobar, O.M. Echeverría and G.H. Vázquez-Nin

Additional information is available at the end of the chapter

1. Introduction

The process of cell death is implicated in several other processes, such as tissue homeostasis, embrionary development, and the elimination of unwanted cells. Programmed cell death is classified first according to the morphological characteristics of the cells observed, and then by the molecular machinery involved in the process. To date, programmed cell death is known to involve apoptosis and autophagy, two processes with different morphological and molecular characteristics.

In mammals, germinal cells are contained in the follicles, specialized structures that develop through several phases of maturation. During follicular growth, cell proliferation and cell death are present simultaneously. During ovarian follicular development, the follicles not selected for the ovulation process are physiologically eliminated. Several studies have shown that in mammalian ovaries follicular atresia is governed by granulosa cell apoptosis (Manabe et al., 2004); however, recent evidence from studies of pre-pubertal (Ortíz et al., 2006; Escobar et al., 2008 and 2010) and adult rats (Escobar et al., 2012) shows that autophagy is an alternative route taken by some germinal cells to induce follicular atresia in the ovary. The emerging importance of autophagy in cellular elimination in the mammalian ovary is a very interesting development.

2. Autophagy as a cell death program

Autophagy is an evolutionary process that eliminates damaged cellular proteins and organelles (Ferraro and Cecconi, 2007). It also plays an important role in bioenergetic management during periods of starvation (Othman et al., 2009), and is the major pathway for the degradation and recycling of intracellular contents.

The autophagy process occurs at a basal level in normal cells under certain adverse conditions, such as starvation, low oxygen levels, and growth factor withdrawal, among others. Under these conditions, autophagy functions as a cytoprotective program that helps maintain cellular homeostasis by recycling the cytoplasmic contents. Another function of autophagy is to eliminate damaged organelles so as to maintain correct cellular functions. Thus, all the features of autophagy in cells perform cytoprotective functions.

In eukaryotic cells, autophagy has been characterized according to the way in which it is carried out: microautophagy, chaperone-mediated autophagy, and macroautophagy (Klionsky, 2006; Massey et al., 2005). In microautophagy, the lysosomal surface directly engulfs the cytoplasm that is to be degraded (Figure 1A). In chaperone-mediated autophagy, the material to be degraded crosses the lysosomal membrane directly (Figure 1B), while macroautophagy, commonly referred to simply as autophagy, is characterized by a double-membrane vesicle (Figure 1C) that encloses (sequesters) organelles and portions of the cytosol (reviewed in Yang and Klionsky, 2009).

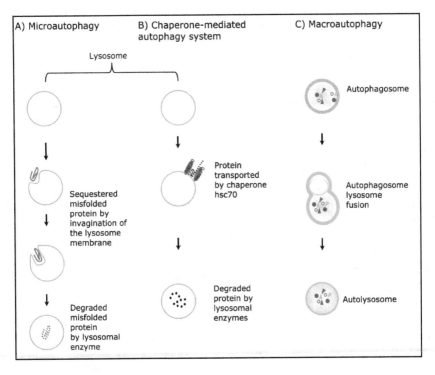

Figure 1. Schematic drawing of autophagic routes. A) Microautophagy: the cytoplasmic contents is directly enclosed by direct invagination from the lysosomal membrane. B) Chaperone-mediated autophagy: the components to be degraded are selectively transported toward the lysosome after interacting with the chaperone hsc70. C) Macroautophagy (commonly called autophagy): the autophagosome containing various cytosolic proteins fuses with lysosomes; subsequently, the contents of the autophagosome is degraded by the lysosomal enzymes.

Morphologically, autophagy is evidenced by the presence of autophagic vesicles, character-ized by a double membrane structure. In mammalian cells, autophagy is initiated by the formation or elongation of the isolation membrane, also called a phagophore. Autophagy entails a sequence of events that includes sequestering the cytoplasmic contents in the double membrane vesicle. Once formed, the autophagosomes are conducted toward lysosomes to constitute the autolysosomes in which the sequestered cellular material is degraded. To avoid degradation itself, the lysosomal membrane is enriched by specific membrane proteins called lysosomal associated membrane proteins (Lamp1 and Lamp2) (Fukuda, 1991).

Though autophagy was first identified in mammalian cells, its molecular characteristics were discovered in yeast. Identification of the participation of the autophagy Atg genes and the subsequent documentation of their homologues in mammals (Yang and Klionsky, 2009) made it possible to determine the molecular machinery involved in the formation and maturation of autophagosomes. TOR kinase is considered an important element in autophagy. When TOR is inhibited under stress conditions, autophagy is induced upon the activation of this kinase; then Atg13 is quickly dephosphorylated, causing a higher affinity for Atg1 and Atg17 that results in an increase in the activity of the Atg1 protein kinase (Kamada et al., 2000; Kabeya et al., 2005). Atg1 kinase plays a pivotal role in controlling autophagy, and its activity is required for the switch from cytoplasm formation to vacuole targeting vesicles (Cvt) and the emergence of autophagosomes (Scott et al., 1996; Matsuura et al., 1997). In mammals, the microtubule-associated protein 1 light chain 3 (LC3) homolog of the Atg8 yeast is an important protein involved in the autophagy process. LC3 is present in autophagosomes and is synthetized in an inactive form called LC3-I, which is later converted into an active membranous form: LC3-II, the lapidated form, which means that it bonds to phosphatidylethanolamine (Wang et al., 2009; Maiuri et al., 2007). LC3 is lapidated via an ubiquitylation-like system that is targeted to the early autophagosome membrane (Kabeya et al., 2004).

During autophagy induction, LC3 is converted from the LC3-I to the LC3-II form. It has been suggested that the amount of LC3-II correlates to the number of autophagosomes present. Autophagy is involved in stress response, developmental remodeling, organelle homeostasis, and disease pathophysiology, and this process may also be used as a host-cell response against bacteria and viruses (reviewed in Kindergaar, 2004).

Additionally, it has been suggested that the effects of autophagy can be either deleteri-ous or protective, depending on the specific cellular context and the stage of the patholog-ical process (reviewed in Rubinsztein et al., 2005). At present, we know that autophagy functions as a form of programmed cell death, classified as type II. One essential differ-ence between physiological autophagy and autophagic cell death is that the levels of autophagy in dying cells are excessive. The role of autophagy as a process of cell death is interesting because it has been observed under certain experimentally manipulated systems. When the pro-apoptotic proteases are inhibited, autophagic levels increase (Yu et al, 2004). Some neurodegenerative diseases, such as Parkinson's disease, have also been associated with the autophagic cell death process (Anglade et al., 1997), and autophagic cell death has been observed as well in remodeling tissues.

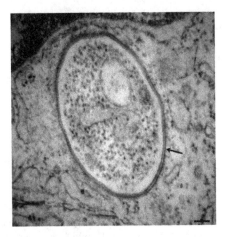

Figure 2. Electron microscopy showing an autophagosome. The arrow points to the two lipid bilayers that surround the cytoplasmatic content. Scale bar 100 nm.

Morphological evidence of autophagy has been found in electron microscopy studies that have shown vesicular structures surrounded by two lipid layers known as autophagosomes (Figure 2). Autophagosomes may contain cytoplasm and cytoplasmic organelles, such as mitochondria and peroxisomes, etc. Autophagy can be evidenced by the immune-microscopic localization of the proteins involved in this process, including LC3/Atg8, or the Lamp1 proteins (Figure 3).

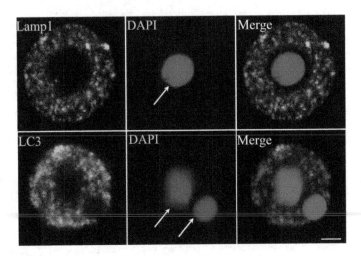

Figure 3. Immunodetection of Lamp1 and LC3. Confocal observations of cellular fractions: DAPI is evident in the nucleus (arrows). The green punctuate fluorescence distinguishes the cytoplasmatic localization of the Lamp1 and LC3 proteins. Scale bar 10 microns.

The role of autophagy as a process of cell death in diverse pathologies, including cancers, has been evaluated widely, but the results from the different studies are somewhat controversial, because at first autophagy functions as a pro-survival strategy, as in the case of tumor cells under certain stimuli; for example, low oxygen or a lack of nutrients (Lefranc et al., 2007). But cancer cells can also use autophagy as a strategy for evading cell death and a means of adapting to an adverse environment. On the other hand, under certain conditions, tumor cells use autophagy as a mechanism of cell death.

3. Follicular atresia

In mammals, the ovary is a paired organ whose principal functions are oocyte production and hormone synthesis. Structurally, the ovary is made up of a medulla and a cortical region where the follicles are generally located. The mammalian ovary is the site of oocyte maturation, which takes place inside a complex structure, the follicle, which is made up of a germinal cell –the oocyte– surrounded by somatic granulosa cells.

Follicles go through several steps before attaining maturation. During this process, various morphological and functional changes occur in the follicle that have led to its development being classified in stages: primordial, primary, secondary, early antral, and antral, according to the number of granulosa layers that surround the oocyte, the size of the follicle, and the presence of the antrum.

Primordial follicles consist of a single flattened cell layer surrounding the oocyte (Figure 4A). In a primary follicle, the granulosa cells around the oocyte acquire a cubical shape in a single-layer cell (Figure 4B). Secondary follicles are characterized by the presence of two or more granulosa cell layers (Figure 5). In this stage, the oocyte increases in size and the granulosa cell layers emerge through intensive proliferation that leads to the formation of the theca interna cell layer (Knight and Glister, 2006). In the secondary follicular phase, the specialized structure associated with the oocyte, called the zona pellucida, is completely discernible. Early-antral and antral follicles (Figure 6) are characterized by the development of a fluid-filled space among the granulosa cells that forms the antral cavity. Granulosa cells are in intercellular contact with neighboring cells via gap junctions (Figure 7), which allow metabolic exchange and the transport of molecules between follicular cells.

During the follicular maturation process, only a few follicles are selected for ovulation, while more than 99% are eliminated via a process denominated follicular atresia (Kaipia and Hsueh, 1977). In ovarian physiology, follicular atresia is a key mechanism for removing the follicles that are not selected for ovulation.

Numerous morphological and biochemical studies have revealed the frequent participation of apoptosis in follicular atresia; indeed, apoptosis came to be considered the cellular route that underlies this process. In caprine ovaries, ultrastructural changes in the granulosa cells show the classic morphological characteristics of apoptosis (Sharma and Bardwaj, 2009). Several pro-apoptotic factors have been identified in granulosa cells, including the FasL-Fas system, TNF-a, and members of the Bcl-2 family of proteins (reviewed in Matsuda et al., 2006). In fact, follicular atresia has been attributed to the alteration of granulosa cells, since studies have demonstrated that these cells synthetize molecules that are essential for follicular maintenance and growth. Furthermore, the death of granulosa cells due to an apoptotic process results in follicular elimination (Matsuda et al., 2012).

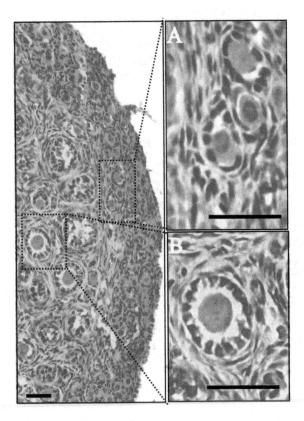

Figure 4. Histological images of a rat ovary. The dotted squares are magnified in the right panel. A) Primordial follicles with flattened pre-granulosa cells. B) Primary follicle with a single layer of cubical granulosa cells. Scale bars 20 microns.

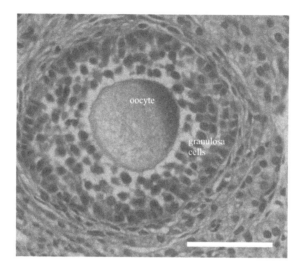

Figure 5. Secondary follicle. The oocyte is surrounded by several layers of granulosa cells. Scale bar 50 microns.

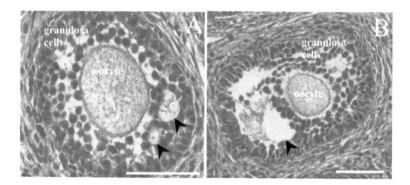

Figure 6. Histological images of a rat ovary. A) Early antral follicle: The arrows show the growing spaces that will form the antral cavity. B) Antral follicle: The arrow points to the antral cavity. Scale bars 50 microns.

4. Granulosa cell death via autophagy

While numerous studies have shown that the process of granulosa cell death is carried out mainly by apoptosis (Feranil et al. 2005; Hurst et al. 2006; Matsuda-Minehata et al. 2006, Lin and Rui, 2010), in some conditions autophagy may be induced in granulosa cells by the process of apoptosis, a process that in rat ovaries is gonadotropin-dependent. These results suggest that both apoptosis and autophagy are gonadotropin-dependent in rat ovaries, and that both processes are involved in regulating granulosa cell death during ovarian follicular development and atresia (Choi et al., 2010). Despite the obvious differences between apoptosis and autophagy, they are now thought to represent points on a continuum of mechanisms of cell death, because the induction of apoptotic cell death is regulated by the process of autophagy. Autophagic cell death is induced by inhibiting the accumulation of autophagosomes in various carcinoma cells, which suggests that the autophagic process may prevent apoptotic death.

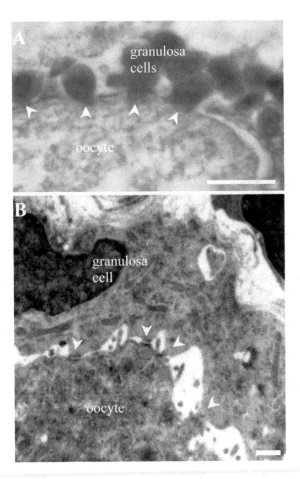

Figure 7. Gap junctions between granulosa cells and the oocyte. A) Optical micrograph showing granulosa cells in strong contact with the oocyte (arrowheads). B) Electron micrograph showing the gap junctions of granulosa cells and the oocyte (arrowheads). Scale Bars: A-50 microns; B-100 nanometers.

In order to investigate the involvement of autophagy in folliculogenesis, and its correlation with apoptosis, isolated rat granulosa cells from immature animals primed with pregnant mare serum gonadotropin were studied. LC3 and autophagic vacuoles were used as markers of autophagy, while cleaved caspase-3 served as the marker of apoptosis. In these conditions, LC3 was expressed by isolated granulosa cells in all developmental stages, and showed a similar expression pattern to the cleaved caspase-3. These results indicate that autophagy is induced in granulosa cells during folliculogenesis in correlation with apoptosis (Choi et al., 2010).

In the human ovary, lectin-like oxidized low-density lipoprotein (LOX) is localized in regressing antral follicles. Treatment with oxLDL (oxidized low-density lipoprotein) causes autophagy in granulosa cells. The process of cell death is characterized by the reorganization of the actin cytoskeleton, abundant vacuoles, autophagosome formation, the absence of apoptotic bodies, and cleaved caspase-3; thus, the reduction of granulosa cells may be mediated by autophagy (Duerrschmidt et al. 2006).

5. Oocyte cell death via autophagy

During the first two trimesters of pregnancy, the number of oocytes in human fetal ovaries increases from approximately 7,200 to 4,933,000 (Mamsen et al., 2011). However, oocyte death begins during the fetal and perinatal stages and continues in newborn, pre-pubertal (Hulas-Stassiak and Gawron, 2011) and adult mammals.

Autophagy is not only a process of cell death; it is also required for cells to survive in conditions of nutrient depletion (Han et al. 2011). Moreover, in murine ovaries it is a cell survival mechanism that maintains the endowment of female gem cells prior to establishing primordial follicle pools (Gawriluk et al. 2011). Several genes have been described as regulators of autophagy; many of them have been conserved from yeast to mammals. In vertebrates, autophagic defects may be lethal if the mutated gene is involved in the early stages of development. However, in different eukaryotes autophagy seems to be crucial during embryogenesis in a way that parallels apoptosis. The earliest autophagic event in mammalian development is observed in fertilized oocytes (Mizushima and Levine, 2010). The identification of *ATG* genes that mediate the initiation and assembly of autophagosomes and their fusion with lysosomes to form autolysosomes brought important advances in our understanding of the various functions of autophagy (Randall-Armant, 2011).

Thus, autophagy seems to be crucial during embryogenesis by acting in tissue remodeling, parallel to apoptosis (Di Bartolomeo et al., 2010). Studies in different organisms indicate that the autophagy pathway in the amoeba *Dictyostelium discoideum* is much more similar to that of mammalian cells than that of *S. cerevisiae*, despite its earlier evolutionary divergence. This indicates that in mammals the autophagic pathway is much older than was previously thought (King, 2012). MicroRNAs are involved in autophagy and are also important regulators of the crosstalk between autophagy and apoptosis (Xu et al. 2012).

ATG genes are also essential for the autophagic pathway in mammalian development (Mizushima and Levine 2010). The oocyte-specific deletion of Atg5, which removes the maternal stores of this protein, produces oocytes that fail to develop past the eight-cell stage, thus demonstrating that autophagy is required during pre-implantation development (Randall Armant 2011). An important increase in the number of autophagosomes takes place immediately after fertilization, which shows the need for autophagosomes after fertilization, in all likelihood to destroy the existing proteins and provide amino-acids for subsequent development (Randall Armant 2011).

In 1-to-28-day-old –i.e., newborn to pre-pubertal– rats, numerous follicles undergo atresia and oocytes are eliminated by processes that include, simultaneously, features of both apoptosis and autophagy. Elements of apoptosis are present in adjacent sections of the same dying oocyte, in the form of active caspase-3 and DNA breaks, as well as large increases of the Lamp1 protein and acid phosphatase, which are present in autophagosomes (Escobar et al., 2008; Escobar et al., 2010). Studies carried out in adult rats have also demonstrated that in all phases of the estrous cycle oocytes die by processes involving features of apoptosis and autophagy simultaneously (Escobar et al., 2012). Morphological changes in atretic oocytes include vacuolization of the cytoplasm, condensation of the mitochondria and segmentation, altera-tions that are not involved in classic apoptosis (Devine et al. 2000). These analyses were carried out using classic markers of apoptosis, such as the TUNEL reaction that reveals DNA frag-mentation, immunolocalization of active caspase-3, and markers of autophagy like a large increase of acid phosphatase, lysosomal hydrolase, and immunodetection of Lamp1, a protein of the lysosomal membrane. These markers are located in the same regions of the oocyte's cytoplasm that present clear vacuoles which correspond to the autophagosomes that became visible using adjacent, semi-thin sections of the same oocyte (Escobar et al., 2008).

In newborn and pre-pubertal spiny mouse oocytes, follicular atresia was studied using markers of apoptosis like the TUNEL reaction, which demonstrate DNA fragmentation and active caspase-3, as well as with markers of autophagy, such as immunodetection of Lamp1. Numerous small clear vacuoles, autophagosomes and Lamp1 staining were found in all follicle types, especially in primordial and primary samples (Figure 8). Active caspase-3 and the TUNEL reaction were detected only in the granulosa cells, showing that both apoptosis and autophagy are involved in follicular atresia, and that these processes are both cell- and developmental-stage specific (Hulas-Stasiak and Gawron, 2011).

Follicular atresia has also been studied in fish ovaries during early and advanced stages of follicular regression. The main events assessed using light microscopy were splits in the zona radiata, yolk degradation and reabsorption, hypertrophy of follicular cells and the accumula-tion of autophagy vacuoles. Labeling for Bcl-2 and cathepsin-D was pronounced in follicular cells when they were involved in yolk phagocytosis. Immunofluorescence for Beclin-1 was significant in the follicular cells that often surround autophagic vacuoles during the advanced stages of follicular regression. TUNEL-positive reactions and immunostaining for Bax and caspase-3 showed the participation of apoptosis in advanced stages of follicular regression. These observations show that both autophagy and apoptosis are activated in some stages of follicular regression in fish ovaries (Morais et al., 2012). Inhibition of the increase of prolifer-ating cell nuclear antigen (PCNA) markedly reduces the apoptosis of oocytes and down-regulates known pro-apoptotic genes, such as Bax, caspase-3, and TNFα, while up-regulating known anti-apoptotic genes like Bcl-2 (Xu et al. 2011).

Retraction of the prolongations of the granulosa cells that normally contacts the surface of the oocyte is one of the early signs of follicular atresia (Devine et al. 2000). Numerous unpublished observations by the authors of this chapter show that the microvilli of the oocyte are elongated after retraction of the prolongations of the granulosa cells during the process of atresia (Figure 9).

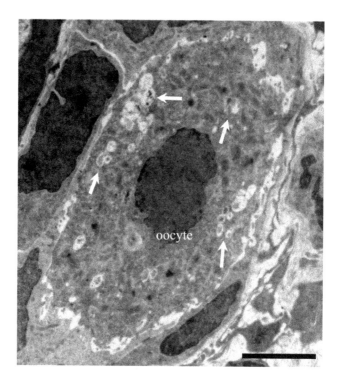

Figure 8. Primordial follicle. The cytoplasm has numerous vacuoles with cytoplasmic contents in different degrees of degradation. Scale bar 2 microns.

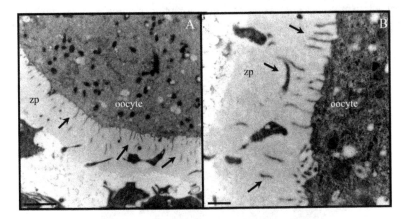

Figure 9. Retraction of the prolongation of the granulosa cells. The arrows point to several microvilli of the oocyte that are elongated after retraction of the prolongations of granulosa cells during atresia. Zp: zona pellucida. Scale bars: A-2 microns; B-500 nanometers.

6. Autophagic cell death in *corpus luteum* regression

The corpus luteum is a transitory ovarian structure formed by cells of the ovulated follicle (figure 10). After an initial proliferation of granulosa cells and closing of the antral cavity, capillaries and theca cells invade the region that was once the granulosa layer of the follicle. During its life span, the corpus luteum undergoes a period of rapid growth that involves hypertrophy, proliferation and differentiation of steroidogenic cells with extensive angiogenesis. After that, it engages in a large production of steroids. Growth factors including insulin-like factor, vascular endothelial growth factor and fibroblast growth factor are important for the development and completion of the dense network of capillaries during the formation of the corpus luteum (Berisha and Schams 2005). There is evidence to suggest that the luteinizing hormone, growth hormones and local regulators such as growth factors, peptides, steroids and prostaglandins are all important regulators of the luteal function. During early corpus luteum development, and up to the mid-luteal stage, oxytocin, prostaglandins and progesterone itself stimulate luteal cell proliferation and functioning, supported by the luteotropic action of several growth factors. High mRNA expression, protein concentration and localization of vascular endothelial growth factor, fibroblast growth factor and members of the family of insulin-like growth factors suggest that they play important roles in the maintenance of the corpus luteum. Progesterone regulates the length of survival of the corpus luteum (Berisha and Shams 2005). In addition, progesterone increases Bcl-2 expression in different stages of the estrous cycle. Treatment of luteal cells with progesterone and prostaglandin PGE2 for 24 hours decreased active caspase-3, while aminoglutethimide, spermine and staurosporine increased caspase-3 activity in luteal cells. These results suggest that progesterone concentrates in luteal cells to protect against apoptosis, while disruption of steroidogenesis and the reduced ability of luteal cells to produce progesterone can induce cell death (Liszewska et al. 2005).

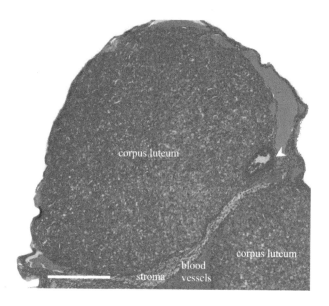

Figure 10. Corpus luteum from a rat ovary. The arrow points to a secondary follicle. Scale bar: 50 microns.

In non-fertile cycles, uterine release of prostaglandin (PG)F(2a) initiates a cascade of events that result in a rapid loss of steroidogenesis and destruction of the luteal tissue (Pate et al. 2012). Periodically, the corpus luteum regresses (luteolysis) and numerous luteal cells undergo cell death processes, mainly through apoptosis and autophagy.

Studies on the role of autophagy in corpus luteum regression have shown an increase of the protein microtubule-associated protein light chain 3 (LC3), a marker of autophagy. Apoptosis was evaluated by measuring cleaved caspase-3 expression (Choi et al. 2011). LC3 expression increases slightly from the early to the mid-luteal stage in steroidogenic cells. The expression levels of the membrane form of LC3 (LC3-II) also increase during luteal stage progression. In the same period, the expression of cleaved caspase-3 also increases. LC3-II expression rises, as do the levels of active caspase-3 in luteal cells cultured with prostaglandin F(2α), which is known to induce corpus luteum regression. These facts suggest that autophagy of luteal cells is directly involved in corpus luteum regression, and correlates with an increase of apoptosis (Choi et al. 2011). When autophagosome degradation by fusion with lysosomes was inhibited using bafilomycine A1 (Baf A1) increased apoptotic cell death. Moreover, inhibition of autophagosome formation using 3-methyladenine decreased apoptosis and cell death, suggesting that the accumulation of autophagosomes induces luteal cell apoptosis. The accumulation of autophagosomes increased apoptotic luteal cell death via an increase in the Bax/Bcl-2 ratio and subsequent caspase activation. Therefore, autophagy plays an important role in regulating apoptotic luteal cell death by controlling the Bax-to-Bcl-2 ratio and the subsequent activation of caspases. These experimental results indicate that autophagy is involved in rat luteal cell death through apoptosis, and that it is most prominent during corpus luteum regression (Choi et al. 2011).

Luteal cell regression during the normal postpartum involution of the corpora lutea is characterized by a large increase in the number of lysosomes and the appearance of numerous double-walled autophagic vacuoles, which become evident under electron microscope cytochemistry (Paavola, 1978).

Compelling evidence indicates that both apoptotic and autophagic cell death programs are involved in corpus luteum regression in primates. Beclin-1, an autophagy-related protein, is involved in the relation between apoptosis and autophagy through interaction with the anti-apoptotic protein Bcl-2. In ovarian follicles, Beclin 1 has been found in the theca layer, but granulosa cells are negative. After ovulation, Beclin-1 is present in theca-lutein and granulosa-lutein cells. The expression of Beclin 1 is related to the functional and structural status of the corpus luteum, as it is a factor in cell survival and plays important roles in the life span of the human corpus luteum (Gaytán et al. 2008).

An endocrine type, voltage-activated sodium channel was identified in the human ovary and human luteinized cells. Whole-cell patch-clamp studies showed that the voltage-activated sodium channels in granulosa cells are functional and tetrodotoxin-sensitive. Luteotrophic hormone was found to decrease the peak amplitude of the sodium current within seconds. Treatment with hGC (human chorionic gonadotropin) for 24-48 hours suppressed not only the mRNA levels in voltage-activated sodium channels, but also the mean sodium peak currents and resting potentials. Tetrodotoxin preserves a highly differentiated cellular phenotype, whereas veratridine not only increases the number of secondary lysosomes but also leads to a reduced progesterone production. In luteinized granulosa cells in culture, abundant secondary lysosomes were evident in the regressing corpus luteum, suggesting a functional link between the voltage-activated sodium channel activity and autophagic cellular regression *in vivo* (Bulling et al., 2000).

Taken together, these data show that several factors are involved in corpus luteum regression. One type of factors includes the process of eliminating the different types of cells that form the corpus luteum, while other types of factors are those involved in destroying the structure of this transitory organ. The normal programmed cell death processes –apoptosis and autophagy– are involved in cell elimination in the corpus luteum. Most authors have found that the most frequent process of cell death is apoptosis; however, very detailed studies demonstrate that both processes are often present simultaneously, as in the case of cell elimination in other organs.

7. Conclusions

Recent years have seen interest grow in the different routes of cell death. Today, two types of programmed cell death are known: apoptosis and autophagy. Cell death in follicle structures is a continuous event during the life of female organisms. Several studies have demonstrated the active participation of apoptosis in this process, but recent biochemical and morphological evidence has revealed the participation of autophagic cell death in oocyte elimination during this physiological process. In granulosa cell death and corpus luteum regression, experimental evidence has shown that autophagy is an active route in the process of cellular elimination. Future studies should test for different stimuli and molecular mechanisms involved in the autophagic cell death process in follicular atresia in vertebrates.

Acknowledgements

The authors would like to thank the grants CONACYT 180526 and PAPIIT IN212912. They also thank Paul C. Kersey Johnson for reviewing the English word usage and grammar.

Author details

M.L. Escobar, O.M. Echeverría and G.H. Vázquez-Nin

Departamento de Biología Celular. Facultad de Ciencias. Universidad Nacional Autónoma de México, Mexico

References

[1] Anglade, P, Vyas, S, Javoy-agid, F, Herrero, M. T, Michel, P. P, Marquez, J, Mouatt-prigent, A, Ruberg, M, Hirsch, E. C, & Agid, Y. (1997). Apoptosis and autophagy in nigral neurons of patients with Parkinson's disease. *Histol Histopathol 12*, 25-31.

[2] Berisha, B, & Schams, D. (2005). Ovarian function in ruminants. *Domest Anim Endocrinol* , 29, 307-317.

[3] Bulling, A, Berg, F. D, Berg, U, Duffy, D. M, Ojeda, S. R, Gratzi, M, & Mayerhofer, A. (2000). Identification of an ovarian voltage-activated Na+ channel type: hints to involvement in luteolysis. *Mol Endocrinol* , 7, 1064-1074.

[4] Choi, J, Jo, M, Lee, E, & Choi, D. (2011). The role of autophagy in corpus luteum regression in the rat. *BiolReproduct 85*, 465-472.

[5] Choi, J. Y, Jo, M, Lee, E. Y, & Choi, D. (2011). Induction of apoptotic cell death via accumulation of autophagosomes in rat granulosa cells. *Fertility and Sterility* , 95, 1482-1485.

[6] Devine, P. J, Payne, C. M, Mccusney, M. K, & Hoyer, P. B. (2000). Ultrastructural evaluation of oocytes during atresia in rat ovarian follicles. *Biol Reprod* , 63, 1245-1252.

[7] Di Bartolomeo SNazio F, Cecconi F. ((2010). The role of autophagy during development in higher eukaryotes. *Traffic* , 10, 1280-1289.

[8] Duerrschmidt, N, Zabirnyk, O, Nowicki, M, Hmeidan, F. A, Blumenauer, V, Burlak, J, & Spanel-borowski, K. (2006). Lectin-like oxidized low-density lipoprotein receptor-1-mediated autophagy in human granulosa cells as an alternative of programmed cell death. *Endocrinology* , 147, 3851-3860.

[9] Escobar, M. L, Echeverría, O. M, Ortiz, R, & Vázquez-nin, G. H. (2008). Combined apoptosis and autophagy, the process that eliminates the oocyte of atretic follicles in immature rats. *Apoptosis* , 13, 1253-1266.

[10] Escobar, M. L, Echeverría, O. M, Sánchez-sánchez, L, Méndez, C, & Pedernera, E. Vázquez-Nin

[11] Escobar Sánchez MLEcheverría Martínez OM, Vázquez-Nin GH. ((2012). Immunohistochemical and ultrastructural visualization of different routes of oocyte elimination in adult rats. *Eur J Histochemistry* , 56, 102-110.

[12] Feranil, J, Isobe, N, & Nakao, T. (2005). Apoptosis in the antral follicles of swamp buffalo and cattle ovary: TUNEL and caspase-3 histochemistry. *Reprod Domest Anim* , 40, 111-116.

[13] Ferraro, E, & Cecconi, F. (2007). Autophagic and apoptotic response to stress signals in mammalian cells. *Arch Biochem Biophys* , 462, 210-219.

[14] Fukuda, M. (1991). Lysosomal membrane glycoproteins. Structure, biosynthesis and intracellular trafficking. *J Biol Chem* , 266, 21327-21330.

[15] Gawriluk, T. R, Hale, A. N, Flews, J. A, Dillon, C. P, Green, D. R, & Rucker, E. B. (2011). Autophagy is a cell survival program for female germ cells in the murine ovary. *Reproduction* , 141, 759-765.

[16] Gaytán, M, Morales, C, Sánchez-criado, J. E, & Gaytán, F. (2008). Immunolocalization of beclin 1, a bcl-2-binding, autophagy-related protein, in human ovary: possible relation to life span of corpus luteum. *Cell Tissue Res* , 331, 509-517.

[17] GH. (2010). Analysis of different cell death processes of prepubertal rat oocytes in vitro. *Apoptosis* 15(4):511-526.

[18] Han, Y. K, Ha, T. K, Lee, S. J, Lee, J. S, & Lee, G. M. (2011). Autophagy and apoptosis of recombinant Chinese hamster ovary cells during fed-batch culture; effect of nutrient supplementation. *Biotechnology Bioeng* , 108, 2182-2192.

[19] Hulas-stasiak, M, & Gawron, A. (2011). Follicular atresia in the prepubertal spiny mouse (*Acomys cahirrinus*) ovary. *Apoptosis* , 10, 967-975.

[20] Hurst, P. R, Mora, J. M, & Fenwick, M. A. (2006). Caspase-3 TUNEL and ultrastructural studies of small follicles in adult human ovarian biopsies. *Hum Reprod* , 21, 1974-1980.

[21] Kabeya, Y, Mizushima, N, Yamamoto, A, Oshitani-okamoto, S, Ohsumi, Y, & Yoshimori, T. and GATE16 localize to autophagosomal membrane depending on form-II formation. *J Cell Sci* , 117, 2805-2812.

[22] Kaipia, A, & Hsueh, A. J. (1997). Regulation of ovarian follicle atresia. *Annu Rev Physiol 59*, 349-363.

[23] Kamada, Y, et al. (2000). Tor-mediated induction of autophagy via an Apg1 protein kinase complex. *J Cell Biol 150*, 1507-1513.

[24] King, J. S. (2012). Autophagy across the eukaryote: Is S. cerevisiae the odd one out? *Autophagy* , 7, 1159-1162.

[25] Kirkegaard, K, Taylor, M. P, & Jackson, W. T. (2004). Cellular autophagy: surrender, avoidance and subversion by microorganisms. *Nat Rev Microbiol* , 2, 301-314.

[26] Klionsky, D. J. (2005). The molecular machinery of autophagy: unanswered questions. *J Cell Sci* , 118, 7-18.

[27] Knight, P. G, & Glister, C. (2006). TGF-β superfamily and ovarian follicle development. *Reproduction* , 132, 191-206.

[28] Lefranc, F, Facchini, V, & Kiss, R. (2007). Proautophagic drugs: A novel means to combat apoptosis-resistant cancers, with a special emphasis on glioblastomas. *Oncologist* , 12, 1395-1403.

[29] Lin, P, & Rui, R. (2010). Effects of follicular size and FSH on granulosa cell apoptosis and atresia in porcine antral follicles. *Mol Reprod Dev 77*(8), 670-678.

[30] Liszewska, E, Rekawiecki, R, & Kotwica, L. (2005). Effect of progesterone on the expression of bax and bcl-2 on caspase activity in bovine luteal cells. *Prostaglandins Other Lipid Mediat 78*(1-4):67-81.

[31] Maiuri, M. C, Zalckvar, E, Kimchi, A, & Kroemer, G. (2007). Self-eating and self-killing: crosstalk between autophagy and apoptosis. *Nat Rev Mol Cell Biol* , 8, 741-752.

[32] Mamsen, L. S, Lutterodt, M. C, Andersen, E. W, Byskov, A. G, & Andersen, C. Y. (2011). Germ cell numbers in human embryonic and fetal gonads during the first two trimesters of pregnancy: analysis of six published studies. *Hum Reprod* , 8, 2140-2145.

[33] Manabe, N, Goto, Y, Matsuda-minehata, F, Inoue, N, Maeda, A, Sugimoto, M, Sakamaki, K, & Miyano, T. (2004). Regulation mechanism of selective atresia in porcine follicles: Regulation of granulosa cell apoptosis during atresia. *J Reprod Dev* , 50, 493-514.

[34] Massey, A. C, Zhang, C, & Cuervo, A. M. (2006). Chaperone-mediated autophagy in aging and disease. *Curr Top Dev Biol* , 73, 205-235.

[35] Matsuda, F, Inoue, N, Manabe, N, & Ohkura, S. (2012). Follicular growth and atresia in mammalian ovaries: regulation by survival and death of granulosa cells. *J Reprod Dev* , 58(1), 44-50.

[36] Matsuda-minehata, F, & Inoue, N. Goto Yasufumi, Manabe N. ((2006). The regulation of ovarian granulosa cell death by pro- and anti-apoptotic molecules. *J Reprod Dev* 52(6), 695-705.

[37] Matsuura, A, Tsukada, M, Wada, Y, & Ohsumi, Y. a novel protein kinase required for the autophagic process in Saccharomyces cerevisiae. Gene , 192, 245-250.

[38] Mcgee, E. A, & Hsueh, A. J. (2000). Initial and cyclic recruitment of ovarian follicles. *Endocr Rev* , 21(2), 200-214.

[39] Mizushima, N, & Levine, B. (2010). Autophagy in mammalian development and differentiation. *Nat Cell Biol* , 12, 823-830.

[40] Morais, R. D, Thomé, R. G, Lemos, F. S, Bazzoli, N, & Rizzo, E. (2012). Autophagy and apoptosis interplay during follicular atresia in fish ovary: a morphological and immunocytochemical study. *Cell Tissue Res* , 347, 467-478.

[41] Ortiz, R, Echeverría, O. M, Salgado, R, Escobar, M. L, & Vázquez-nin, G. H. (2006). Fine structural analysis of the processes of cell death of oocytes in atretic follicles in new born and prepubertal rats. *Apoptosis* , 11, 25-37.

[42] Othman, E. Q, Kaur, G, Mutee, A. F, Muhammad, T. S, & Tan, M. L. (2009). Immuno-histochemical expression of MAP1LC3A and MAP1LC3B protein in breast carcinoma tissues. *J Clin Lab Anal* 23(4), 249-258.

[43] Paavola, L. G. (1978). The corpus luteum of guinea pig. III. Cytochemical studies on the Golgi complex and GERL during normal postpartum regression of luteal cells, emphasizing the origin off lysosomes and autophagic vacuoles. *J Cell Biol* , 79, 59-73.

[44] Pate, J, Johnson-larson, C, & Ottobre, J. (2012). Life and death in the corpus luteum. *Reprod Domest Anim* , 47, 297-303.

[45] Randall Armant D(2011). Autophagy's expanding role in development: implantation is next. *Endocrinology* , 152, 11739-11741.

[46] Rubinsztein, D. C. DiFiglia M, Heintz N, Nixon RA, Qin ZH, Ravikumar B, Stefanis L, Tolkovsky A. ((2005). Autophagy and its possible roles in nervous system diseases, damage and repair. *Autophagy* , 1, 11-22.

[47] Scott, S. V, Hefner-gravink, A, Morano, K. A, Noda, T, Ohsumi, Y, & Klionsky, D. J. (1996). Cytoplasm-tovacuole targeting and autophagy employ the same machinery to deliver proteins to the yeast vacuole. *Proc Natl Acad Sci USA* , 93, 12304-12308.

[48] Wang, A. L, & Boulton, M. E. Dunn WA Jr, Rao HV, Cai J, Lukas TJ, Neufeld AH. ((2009). Using LC3 to monitor autophagy flux in the retinal pigment epithelium. *Autophagy 5*, 1190-1193.

[49] Xu, B, Hua, J, Zhang, Y, Jiang, X, Zhang, H, Ma, T, Zheng, W, Sun, R, Shen, W, Sha, J, Cooke, H. J, & Shi, Q. (2011). Proliferating cell nuclear antigen (PCNA) regulates primordial follicle assembly by promoting apoptosis of oocytes in fetal and neonatal mouse ovaries. *Plos ONE* 6(1):e16046.

[50] Xu, J, Wang, Y, Tan, X, & Jing, H. (2012). Micro RNAs in autophagy and their emerging roles in crosstalk with apoptosis. *Autophagy* , 8(6), 873-882.

[51] Yang, Z. Klionsky DJ: An overview of the molecular mechanism of autophagy. *Curr Top Microbiol Immunol* , 335, 1-32.

[52] Yu, L, Alva, A, Su, H, Dutt, P, Freundt, E, Welsh, S, Baehrecke, E. H, & Lenardo, M. J. (2004). Regulation of an ATG7-beclin 1 program of autophagic cell death by caspase-8. *Science* , 304, 1500-1502.

Chapter 9

Integrin and Adhesion Regulation of Autophagy and Mitophagy

Eric A. Nollet and Cindy K. Miranti

Additional information is available at the end of the chapter

1. Introduction

Cell differentiation is a dynamic process that generates a functionally distinct cell from its progenitor. For example, human erythrocytes lack most organelles - including a nucleus - while erythrocyte precursors have a complete set of organelles. Autophagy plays a critical role in organelle elimination in differentiating erythrocytes. On the other hand, most other differentiated cells in the body do not lose their organelles, and would not seem to heavily depend on autophagy during differentiation. However, evidence indicates that these cells require regulated autophagy during the differentiation process. For example, during keratinocyte differentiation a basal cell detaches from the basement membrane and is pushed to the upper strata, and as the cell ascends its intracellular contents are replaced with copious amounts of keratin. Drastic changes such these require cells to eliminate or turn over a large amount of biomass. Differentiation also causes a shift in signaling and survival pathways. An example of this is the prostate gland where the luminal cells require PI3K for survival while the undifferentiated basal cells do not; they depend on MAPK signaling for their survival. As we will discuss below, several different cell types use and require the autophagic pathway to properly differentiate and survive. While links between autophagy and differentiation are rapidly being identified, the mechanisms that trigger autophagic processes during differentiation are poorly understood. Genetics has served as a powerful tool for identifying the components of the autophagy machinery, but how they are integrated with cellular cues that trigger differentiation need further characterization. We will discuss the recently identified link between autophagy and cell adhesion, and its role in cellular differentiation and survival.

2. Role of autophagy during differentiation

2.1. Erythrocytes

Erythrocytes are especially unique cells; they lack a nucleus, internal membrane-bound organelles, and ribosomes, and they are packed with the oxygen transporter hemoglobin. The cell arises from reticulocytes, a nucleated progenitor capable of generating the needed surplus hemoglobin and the erythrocyte itself [1,2]. Generating such a simple cell requires processes that eliminate organelles not essential for fully differentiated erythrocytes. A series of publications from 2008 to present not only elucidated the involvement of autophagy during reticulocyte differentiation, but also discovered that the mitochondria are specifically targeted for autophagy by a protein called Nix [3-7].

Nix and its related family member, Bnip3, are unique mitochondrial-localized BH3-only proteins. Although they can induce apoptosis when over expressed like most of the BH3-only proteins, Nix and Bnip3 function to stimulate autophagy and mitophagy as well. The mitophagic function of Nix was discovered in differentiating reticulocytes. Researchers knew that Nix expression dramatically increases in the terminal stage of reticulocyte differentiation, but the purpose for this remained unknown for several years [8]. In a relatively simple experiment, researchers harvested erythrocytes from Nix$^{-/-}$ mice and using mitotrophic dyes quantified mitochondria-containing erythrocytes. A significant population of the Nix$^{-/-}$ erythrocytes still contained mitochondria. The absence of Nix did not affect LC3 levels, and autophagosomes were still present. However, the autophagosomes in Nix$^{-/-}$ erythrocytes contained significantly less mitochondria than the wild type controls [4].

Erythrocytes from Ulk1$^{-/-}$ mice retained mitochondria and ribosomes, indicating the necessity of the general autophagy program in reticulocyte differentiation. Inhibiting autophagy with the class III-PI3K inhibitor 3-methyladenine (3-MA) in wild type reticulocytes prevented mitochondrial clearance as well. The peripheral blood in Ulk$^{-/-}$ mice contained an increased number of reticulocytes indicating a reduced capacity to be converted to erythrocytes [3]. Transplantation of fetal liver cells from Atg7$^{-/-}$ mice into irradiated wild type mice resulted in overall fewer erythrocytes, and these erythrocytes contained mitochondria.[7]. Thus, both general macroautophagy, and organelle-specific autophagy is required for erythrocyte differentiation. The above studies focused on mitochondria, but whether specific autophagic targeting of other organelles, such as ribosomes and ER, is mediated by Nix or other organelle-specific factors during reticulocyte differentiation remains unanswered. Bnip3, like Nix, localizes to the ER, and researchers recently demonstrated that Bnip3 targets the ER for autophagy [9]. This raises the possibility that Nix may also target the ER for autophagy. Targeted autophagic degradation of ribosomes, termed ribophagy, occurs in S. cerevisiae. However, whether selective ribophagy occurs in higher eukaryotes remains unknown [10]. These questions would be interesting to answer in the reticulocyte differentiation model.

2.2. Lymphocytes

In the same Atg7$^{-/-}$ transplantation model, there was a four-fold reduction in the number of white blood cells and nine-fold reduction of lymphoid cells [7]. This observation was further validated in hematopoietic-specific stem, fetal, and adult cell Atg7 knockout mouse models [9, 10, 11]. Mitochondrial accumulation was observed in both CD8$^+$ and CD4$^+$ T-cells, and these cells succumbed to apoptosis in vitro more readily than the wild type controls [11]. T-cell specific knockouts of Atg5 and Atg7 also displayed increased amounts of mitochondria and were more susceptible to apoptosis [12,13]. Considering that the overall population of mature T lymphocytes decreased and sensitivity to apoptosis increased in the autophagy-deficient T lymphocyte models, it seems that the lymphopenia occurs due to excess cell death. Puo et al tested this by stimulating T-cell proliferation in wild type and Atg5$^{-/-}$ T-cells and quantifying the amount of daughter cells produced by using the stable dye, 5,6-carboxyfluorescein diacetate succinimidyl ester (CFSE). CFSE diffuses into both cells during mitosis, causing a decrease in individual cell fluorescent intensity with every division. The analysis demonstrated that stimulated Atg5$^{-/-}$ T-cells did not produce daughter cells after three days and the authors concluded that Atg5 is necessary for T-cell proliferation. However, CFSE cannot differentiate between a dead or permeabilized cell and a non-proliferating cell, and over the three day period there was a moderate increase in apoptosis [14,15]. In this model, it appears that autophagy is required during T-cell hematopoiesis, at least in part, to prevent death of the newly differentiated cells. It would be interesting to investigate this in an inducible model, to test whether the dependence on autophagy is transient, i.e. only during early differentiation stages, or required to maintain the differentiated population.

2.3. Adipocytes

Evidence also supports a role for autophagy in adipocyte differentiation. Treatment of mouse embryonic fibroblasts (MEFs) with a cocktail of differentiation-inducing factors over the course of two weeks causes accumulation of lipid analogous to mature adipocytes. In the absence of Atg5 the MEFs fail to accumulate lipid droplets or display the molecular or structural phenotype of differentiated adipocytes. Furthermore, Atg5$^{-/-}$ mouse pups have far fewer adipocytes [16]. Knocking down Atg7 in 3T3-L1 preadipocytes attenuated the accumulation of triglycerides and several markers of mature adipocytes as well. This was further validated in vivo in an adipose-specific Atg7 knock out mouse; the total fat in the knock out mouse weighed less than half that of the wild type mice. Interestingly, the adipose-specific Atg7$^{-/-}$ mouse had increased amounts of brown adipose tissue (BAT) and less white adipose tissue (WAT) relative to the wild type mouse, and both the lipid droplets and adipocytes were smaller and more densely packed than normal adipocytes [17,18]. The brown appearance of BAT is due to high quantities of densely packed mitochondria. BAT can uncouple the mitochondrial ATPase and continue oxidizing nutrients to generate thermal energy in place of the chemical energy provided by ATP. Zhang et al. validated these results, but went further and confirmed an increased presence of mitochondria [18]. Thus, while autophagy deficient adipocytes fail to express differentiation markers and lack lipid droplet morphology, some of the effects seen in autophagy-deficient adipocytes may result from the accumulation of mitochondria. Indeed, increased lipid catabolism in the form of β-oxidation occurred in the Atg7[-/-] adipocytes [17]. These findings points to a potentially functionally significant role for mitophagy during adipocyte differentiation. WAT primarily stores fatty acids for use under starvation conditions. Perhaps mitophagy is a necessary step for preventing wasteful energy expenditure through lipid catabolism in adipocytes. However, loss of the pro-mitophagic protein Bnip3 causes hepatocytes to accumulate mitochondria like adipocytes, but at the same time actually increase anabolic processes such as fatty acid synthesis [19]. It would be interesting to identify the cause of these differences.

2.4. Epithelial cells

Autophagy plays a role in different aspects of epithelial differentiation. One aspect resembles that seen in erythrocytes. Keratinocyte basal cells begin differentiation by detaching from the basement membrane and moving apically toward their fate of cornification and exfoliation. In the granulation stage, keratinocytes lose their organelles and nucleus [20]. Increased proteasome activity accompanies granulation and accounts for some of the degradative action, but entire organelles may require the autophagy pathway. Granulating keratinocytes are packed with lysosome bodies, and examination of expression patterns during differentiation revealed increased expression of the pro-autophagic proteins Beclin and Sirt1 [21,22]. Whether these events are more specifically controlled by organelle-specific autophagy mechanisms as is observed in erythrocytes has yet to be determined.

A question that remains is whether there is a role for autophagy in earlier epithelial differentiation stages, such as during the first emergence of suprabasal cells in striated epithelium. In one study, 3-MA-treated keratinocytes failed to express the differentiation marker involucrin when stimulated to differentiate in low glucose and high calcium [22]. However, this study should be interpreted with caution, because the reduction in glucose used to induce differentiation is a metabolic stressor. Because the metabolic state of cells regulates rates of autophagy, the increase in autophagy seen here may be an independent coincidence caused by the metabolic stress used to induce differentiation. Additionally, the sole use of 3-MA to inhibit autophagy brings up the question of selectivity. 3-MA also inhibits class I PI3K [23], an important regulator of keratinocyte differentiation [24,25].

The most striking link between autophagy and early epithelial differentiation comes from studies in the mammary gland. Autophagy promotes the survival of differentiated breast epithelial cells in 3D models of breast acinar formation [26]. Like keratinocytes, the breast epithelial cells lose adhesion to the ECM and push away from the basement membrane as differentiation occurs. Autophagy is increased in the differentiating cells in the center of the acini. However, inhibiting autophagy with 3-MA during differentiation increases luminal cell caspase-3 cleavage and death rather than blocking differentiation per se [27]. This suggests autophagy controls the rate of lumen cell loss, but the reason for needing this intermediate step is not clear.

Another example of autophagy in epithelial differentiation is in development of the embryoid body (EB). Embryoid bodies develop spontaneously from cultured embryonic stem cells, forming a spherical body of cells that can differentiate into various cell types [28]. The inner cells of the EB display increased amounts of autophagic vacuoles similar to the breast epithelia 3D acini. However, unlike what happens in the breast acini, autophagy inhibition resulted in the accumulation of dead cell bodies in the EB lumen, suggesting that in this case increased autophagy is responsible for clearing out the dead cell debris accumulating at the center of the EB spheroid. In EB, cavitation occurs in response to increased hypoxia. Qi et al concluded that the hypoxic environment increased apoptosis inducing factor (AIF), which in turn increased ROS production and HIF2α signaling, and subsequently up-regulated pro-autophagic Bnip3. Knocking down any of the aforementioned proteins delayed cavitation of the EB. However, unlike the 3D breast acini model, inhibiting autophagy delayed caspase cleavage [29].

The examples in breast and embryoid bodies indicate that one of the roles of autophagy is to control when and how cell death occurs during differentiation, whereas autophagy is required for organelle clearing and survival in granulocytic keratinocytes and erythrocytes [21]. Thus, depending on the situation or cell state, autophagy can play distinct roles in differentiation.

A key question that remains in many of these models is why autophagy is important during differentiation. Mitochondrial clearance is likely important for the functional aspect of erythrocytes; if erythrocytes still contained mitochondria and carried out oxidative phosphorylation they would probably deplete much of their own oxygen in circulation before reaching tissues needing oxygen. Failure to eliminate mitochondria during adipocyte differentiation results in increased catabolism and a poor ability to store fat. However, in many systems autophagy seems to allow the cells to simply survive differentiation. Perhaps the metabolic stress of differentiating takes a toll on the integrity of the mitochondria and other intracellular components and autophagy increases just to keep up with the increased demand for maintenance. On the other hand, autophagy may have specific tumor suppressive effects during differentiation. Loss or reduction in autophagy appears to be required for tumor initiation and it is well established that tumorigenesis is linked with aberrant differentiation. One possibility is that autophagy actually promotes terminal differentiation, creating an antagonist to tumorigenesis. The sacrifice the tumor cells make is to create a greater dependence on anti-apoptotic survival pathways to compensate for the lack of the survival benefit of autophagy. However, as cancer progresses, some aspects of the autophagy pathway re-emerge as a mechanism to escape death under stress, but this must be balanced by preventing differentiation and reduced dependence on mitochondria for energy.

3. Adhesion and regulation of autophagy

In the case of multilayered epithelia, such as skin, bladder, prostate, and breast, differentiation occurs as the cells detach from the basement membrane and ascend apically. Such major changes require extensive reprogramming of signaling networks and gene expression, and cells must eliminate or inhibit the former cellular programming machinery. Failure to efficiently modify the programming can cause cellular stress and/or oncogenesis. Perhaps an immediate increase in autophagy aids in temporal separation of the undifferentiated and differentiated cell signaling and programming pathways. Coincident with this process is the need to maintain cell survival. This brings to question what signaling molecules in differentiation regulate autophagy.

3.1. Integrins and detachment-induced autophagy

Integrins, a family of heterodimeric transmembrane proteins, mediate cell adhesion to the extracellular matrix (ECM). Integrins regulate a variety of functions at the cellular and molecular level [30]. In stratified epithelial cell differentiation the integrins holding basal cells to the basement membrane become internalized and eliminated as the cell stratifies. Integrins form focal points with complexes of signaling molecules to maintain the connection between the matrix and the contracting cytoskeleton [31]. Cell survival is a key function of integrins. Loss of cell adhesion due to disengagement of integrins from their ECM ligand simultaneously activates extrinsic and intrinsic apoptotic cell death termed anoikis [32]. This signaling pathway prevents exfoliated or damaged cells from surviving detachment and adhering at an improper location.

Detachment and anoikis also aids in lumen clearing in secretory epithelium. As epithelial cells become contact inhibited, they force some cells off the basement membrane [33]. In undifferentiated epithelium this loss of adhesion would normally trigger anoikis and apoptosis. However, in differentiating epithelium, autophagy is triggered which delays the rapid onset of anoikis. This was demonstrated in both 3D acini and in suspended cells. Autophagy was induced in a subpopulation of human mammary epithelial MCF10A and canine kidney epithelial cells when they were placed in suspension, which allowed them to survive anoikis. Inhibiting autophagy by knocking down the autophagy proteins ATG5, ATG6, and ATG7 in suspended MCF-10A cells increased cleaved caspase-3 positive cells and decreased replating efficiency [27]. Further studies demonstrated that the decision to undergo autophagy, as opposed to anoikis, is controlled in part by up-regulation PERK to suppress ROS, through activation of the ER stress pathway, and resulting in increased expression of ATG6 and ATG8 [34]. When squamous cell carcinomas (SCCs) were isolated from FAK-/- mice (which presumably have altered integrin-based adhesion), or when SCCs isolated from wild type mice were placed in suspension, the rate of autophagy was significantly increased. This resulted in the targeting of Src to autophagosomes through c-Cbl in an E3-ligase-independent mechanism [35]. The major conclusion drawn from these studies is that loss of matrix adhesion induces autophagy.

3.2. Attachment-induced autophagy

This latter conclusion is partially contradicted by several studies demonstrating that adhesion to matrix is required to promote autophagy. In a study on human primary basal prostate epithelial cells (PrEC) researchers found that blocking integrin interactions with the ECM inhibited autophagy induced by starvation. In culture, PrECs secrete and adhere to a laminin 5, via integrins α3β1 and α6β4. These integrins were necessary for maintaining cell survival through Src and ligand-independent EGFR activation. Blocking integrin α3, α6, or β4 with antibodies inhibited both survival and autophagy in these cells. Particularly interesting is that inhibiting autophagy with 3-MA or blocking integrin α3 antibodies induced caspase cleavage, but inhibiting EGFR alone did not. Since a class I-specific inhibitor could not induce apoptosis in PrECs, it is unlikely that off target inhibition of class I PI3K by 3-MA resulted in caspase cleavage. In addition, cell adhesion or inhibition of EGFR had no effect on Akt phosphorylation, which suggests that integrin-mediated maintenance of autophagy does not occur through the PI3K/Akt pathway in PrECs [36]. This study demonstrated that integrin adhesion mediates autophagy. While the exact downstream signaling pathway from integrins to autophagy remains unclear, survival was dependent on Erk signaling which is known to promote autophagy. A subsequent study demonstrated that in prostate cancer cells the androgen receptor promoted survival via integrin α6β1 activation of IKK/NF-κB signaling, independent of the PI3K/Akt pathway [37]. A similar dependence on the IKK/NF-κB pathway for survival is mediated by integrin α6β4 in polarized cells of 3D breast acinar structures and over activation of this pathway is sufficient to overcome suspension-induced anoikis and apoptosis [38-40]. Given that activation of the IKK/NF-κB pathway can induce autophagy [41], it is reasonable to suspect that integrins may regulate autophagy in part through the IKK/NF-κB pathway (Figure 1A). Further insight into how integrins regulate NF-κB and autophagy is needed.

Stimulation of smooth muscle cells with the integrin ligand osteopontin, stimulated autophagy related-genes, autophagosome formation, and ultimately cell death [42]. This effect was mediated by p38-MAPK through integrin and CD44 (Figure 1C). In a Drosophila genetic screen, paxillin, a downstream target of integrins, was found to associate with Atg1 and to be required for starvation-induced autophagy in MEFs [43]. However, in the fly, the ability of paxillin to facilitate autophagy was not linked with integrins. Whether this is true in the mammalian model has not been determined. Thus, several studies implicate integrin-based adhesion in both suppressing as well as stimulating autophagy. Since all these studies were conducted in different model systems and under different culture conditions, it is difficult to draw a definitive conclusion. Therefore, when studying different biological events it is necessary to consider several possible mechanisms and signaling pathways that could influence autophagy.

Some consideration should be given to a potential alternative mechanism to explain detachment-induced autophagy. Although it was quite thoroughly demonstrated that the cells increased the rate of autophagy after losing ECM contact, in both the EB and acini models the cells are undergoing a differentiation program and remaining in contact with neighboring cells. In the MCF-10A breast model where cells were put into suspension and 25% survived anoikis, these cells existed in large aggregates [27]. In light of the fact that increased cell density also increases autophagy in vitro, one could hypothesize that the proteins mediating cell-cell contact may be signaling to increase autophagy in the detached cells [23].

3.3. Cell-cell adhesion-induced autophagy

A relatively new process called entosis, whereby cell-cell adhesion induces the entry of one cell into a neighboring cell, was found to trigger an autophagic response [44] by mechanisms that resemble pathogen engulfment by immune cells. The ability of cells to entose depends on adherens junctions in the absence of integrin adhesion [45]. The major protein that mediates

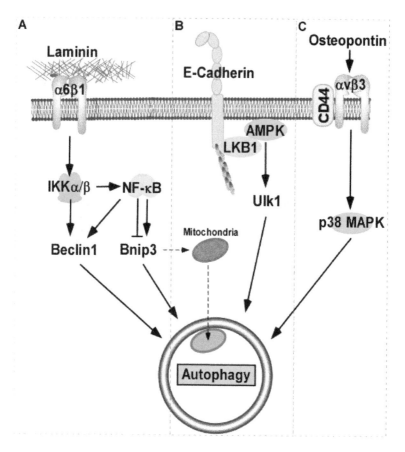

Figure 1. Cell Adhesion Pathways that Promote Autophagy. (A) Integrin α6β1 activates NF-κB through IKK signaling. Together, IKK and NF-κB can up-regulate and activate Beclin, and NF-κB in turn regulates pro-mitophagic Bnip3 expression. (B) AMPK localizes with LKB1 at adherens junctions where LKB1 activates AMPK signaling, which can activate autophagy through Ulk1. (C) Osteopontin activates p38 MAPK-induced autophagy through CD44 and αvβ3.

epithelial cell-cell contact through adherens junctions is E-cadherin. This vital protein holds the epithelium together and promotes the survival of cells [46-48]. Adherens junctions mediate the connection between cells by essentially bridging the actin cytoskeleton of two different cells. A plethora of signaling events are controlled by E-cadherin. In the case of autophagy, one point of interest is regulation of AMPK.

In response to low ATP levels, AMPK directly phosphorylates and activates Ulk1 in parallel to deactivating the autophagy suppressor mTOR [49]. Membrane bound E-cadherin localizes with the AMPK activator LKB1 (Figure 1B), and by doing so increases AMPK activation by bringing it into close proximity with LKB1 [50]. In one study AMPK was required for the drug

Atorva to induce an endoplasmic reticulum stress response, autophagy, and phosphorylation of eIF2α [51]. Interestingly, in the MCF-10A detachment-induced autophagy model, suspended cells had increased levels of activated PERK, a well-documented cytoprotective ER stress-response protein that induces autophagy, and increased phosphorylation of the downstream PERK target eIF2α in conjunction with increased autophagic flux [34]. Considering the overlapping pathways and the correlation of detachment-induced autophagy with cell-cell contact this could be a promising area to investigate.

3.4. Genetic evidence for integrin adhesion and autophagy

Deficiency of the laminin α2 chain (LAMA2) in laminin 2 causes a form of congenital muscular dystrophy known as MDC1A. MDC1A patients present with severe muscle weakness, peripheral neuropathy, and joint contractures [52]. The LAMA2-deficient mouse model (dy^{3K}/dy^{3K}) develops muscle weakness two weeks after birth and dies by 5 weeks with severe muscular dystrophy [53,54]. The dy^{3K}/dy^{3K} mice feature increases in myocyte apoptosis, degeneration/regeneration cycles, variable fiber sizes, and connective tissue hyperproliferation. dy^{3K}/dy^{3K} mice exhibit significantly increased expression of the autophagy related genes Bnip3, Bnip3l, p62, LC3B, GABARAP1, Vps34, Atg4b, Cathepsin L, Lamp2a, and Beclin. Administration of 3-MA to the dy^{3K}/dy^{3K} mice increased lifespans, increased average muscle fiber diameter, decreased the presence of caspase-3 in myocytes, and partially restored muscular morphology. In addition, immunofluorescent staining of muscle biopsies from MDC1A patients showed an increased amount of LC3B in the myocytes. Thus, patients with MDC1A may essentially suffer from excessive muscular autophagy due to a lack of interaction with the matrix protein laminin 2 (Figure 2A). This example supports a role for matrix detachment in inducing autophagy.

On the other hand, one of the LAMA2 receptors, dystroglycan which binds laminin extracellularly and dystrophin intracellularly, may promote adhesion-dependent autophagy. One function of the dystroglycan/dystrophin complex is to bind F-actin to anchor the cytoskeleton in place. Like the dy^{3K}/dy^{3K} mice, the dystrophin mutant (mdx) mice display a muscular dystrophy phenotype. However, in this case, the mdx mice accumulate damaged mitochondria, and inducing autophagy through AMPK activation in these mice ameliorates the disease phenotype, indicating there is a defect in muscle autophagy in mdx mice. This further suggests that adhesion to laminin through the dystroglycan/dystrophin connection is required to maintain autophagy in myocytes [55,56]. It is worth noting that mutations in integrin α7 are associated with muscular dystrophy as well. The integrin dimer α7β1 is another receptor that binds laminin-α2 and resides on myocytes [57]. However, whether α7β1 interactions affect autophagy in these diseases remains unknown. Further evidence supporting the concept that cell adhesion mediates autophagy comes from the study of Ulrich's Congenital Muscular Dystrophy (UCMD) and Bethlem Myopathy. Both diseases involve chronic weakening and degradation of skeletal muscle due to the spontaneous death of myocytes, which ultimately leads to death. Genetic studies discerned that mutations in the gene coding for collagen VI correlate with the condition [58]. Furthermore, genetically modified mice lacking the collagen VI gene (Col6-/-) develop myopathies similar to UCMD and Bethlem Myopathy. The Col6-/-

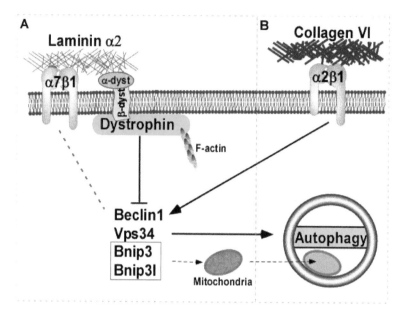

Figure 2. Matrix Adhesion Pathways and Autophagy in Muscle Cells. (A) Laminin α2 is bound by the α- and β-dystro-glycan complex, which in turn binds dystrophin, a protein that anchors the complex onto F-actin. Loss of Laminin α2 caused increase expression of Beclin, Bnip3, Bnip3l, Vps34, and several other autophagy related proteins. Both Bnip3 and Bnip3l can induce mitophagy as well. Integrin α7β1 binds laminin α2 as well, but a connection with autophagy has not been established. (B) Loss of Collagen VI increased pro autophagic and pro mitophagic genes and caused de-creased autophagy along with decreased expression of Beclin and Bnip3. The collagen receptor integrin α2β1 may be responsible for this.

mice have dysmorphic and dysfunctional myofibers characterized by increased apoptosis and reduced muscle strength compared to wild type mice [59]. Electron microscopy of Col6$^{-/-}$ myocytes revealed damaged and misshapen mitochondria and biochemical assays demon-strated that the mitochondria are dysfunctional. Inhibiting mitochondrial permeablization with Cyclosporine A in Col6$^{-/-}$ mice and in patients with homologous myopathies ameliorated the diseases effects [60], and genetic ablation of the cyclosporine A target, Cyclophilin D, also mitigated the symptoms in Col6$^{-/-}$ mice [61]

Later research found that Col6$^{-/-}$ mice muscle tissue expressed lower levels of the pro-auto-phagic proteins Beclin and Bnip3 and displayed a significant decrease in autophagy (Figure 2B). Forced expression of Beclin and Bnip3 partially reversed the disease phenotype in Col6$^{-/-}$ myocytes, and inducing autophagy by fasting the Col6$^{-/-}$ mice reestablished autophagy and decreased apoptosis in the myocytes. Furthermore, fasting or treating the Col6$^{-/-}$ mice with Rapamycin, to block the autophagy inhibitor mTor, decreased the presence of dysfunctional mitochondria [62]. These latter studies on Col6-related myopathies indicate that myocyte cell adhesion to collagen maintains autophagic flux and cell survival, while unregulated detach-ment inhibits autophagy.

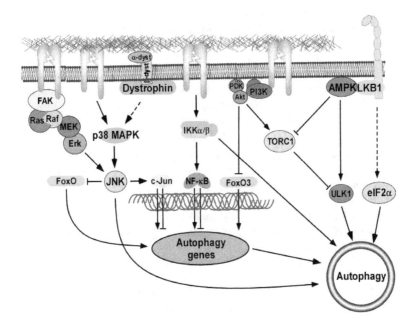

Figure 3. Mechanisms for Cell Adhesion-Induced Autophagy. The intracellular domain of integrins binds to FAK, which activates the Ras/Raf/MEK/Erk signaling cascade. Integrins and dystrophin both can activate p38 MAPK. Both Erk and p38 can phosphorylate JNK and induce autophagy. JNK in turn can either inhibit autophagy through suppression of FoxO or promote autophagy through activation of c-Jun, which can positively and negatively regulate autophagy. Integrins can also activate the IKKα/β–NF-κB pathway. IKKα/β induces autophagy both dependently and independently of NF-κB; however, NF-κB can also inhibit autophagy in some cases. Integrins also activate the PI3K/Akt/TORC1, which inhibit autophagy by phosphorylating and inhibiting FoxO3 and Ulk1. LKB1 binds to E-cadherin and activates AMPK. AMPK both activates Ulk1 and inhibits TORC1, causing an increase in autophagy. Some evidence indicates that cell-cell adhesion may also activate the autophagy-inducer eIF2α.

3.5. Cell adhesion signaling pathways that regulate autophagy

Adhesion proteins interact with a host of proteins and activate several well characterized pathways that affect autophagy. Recall that the dystrophin glycoprotein complex promotes adhesion dependent signaling in myocytes, and the importance of this was highlighted in the Mdx and dy^{3K}/dy^{3K} mice (Figure 2A). When the dystrophin glycoprotein complex is bound to laminin, an intracellular component of the complex called syntrophin binds and activates the Rac1 signaling complex, leading to activation of JNK, a protein that can both positively and negatively regulate autophagy [63] (Figure 3). JNK can induce autophagy by phosphorylating Bcl-2 and thus preventing Bcl-2 from inhibiting Beclin [64]. JNK also regulates several transcription factors that in turn mediate expression or repression of autophagy related genes. For example, genetic ablation of all three JNK isoforms results in constitutive activation of the transcription factor FoxO, causing an increase in autophagy and the effector proteins Bnip3 and Beclin. JNK can also activate the transcription factor c-Jun, a protein which when overexpressed can inhibit autophagy under starvation conditions [65]. On the other hand, another study showed that JNK was required for ceramide induced autophagy and expression of Beclin [66]. Furthermore, Jnk1 can phosphorylate Bcl-2 to release Beclin and promote autophagy [67]. Integrins may also regulate autophagy via p38 MAPK. Osteopontin binds integrins and CD44 and induces autophagy through activation of p38 MAPK [42]. Although some evidence indicates that p38 inhibits autophagy, p38 can activate JNK [68], which could provide a potential mechanism for both positive and negative regulation of autophagy by integrins. Finally, integrins stimulate Erk1/2 signaling as well, which activates autophagy through numerous pathways including activation of JNK [68].

Signaling through IKKα/β–NF-κB is another node for promoting autophagy (Figure 3). IKKα/β regulates the expression of autophagy-related genes by activating NF-κB, which in turn has been shown to both positively and negatively regulate autophagy [41,69,70]. IKKα/β can also activate autophagy independently of NF-κB [70]. Cells expressing constitutively active subunits of the IKK complex had increased autophagy and increased levels of phosphorylated active AMPK and JNK [71]. Considering that blocking integrin adhesion in PrECs subsequently attenuated autophagy [36], and that integrin mediated adhesion to laminin increased IKKα/β - NF-κB signaling enough to significantly increase survival in a prostate cancer line [46], signaling through IKKα/β - NF-κB may be a mechanism by which integrin-mediated-cell-adhesion promotes autophagy. It would be interesting to determine whether this mechanism occurs and whether it is dependent or independent of NF-κB.

PI3K/Akt signaling potently suppresses autophagy [72,73] (Figure 3). Akt activates the TORC1 complex, which in turn phosphorylates and inactivates the autophagy-initiating kinase Ulk1 [73]. Akt itself phosphorylates and inactivates the FoxO3a transcription factor. This prevents FoxO3a from inducing the expression of numerous autophagy- and Mitophagy-related genes [74]. Integrins can activate PI3K/Akt signaling [30], making it possible that integrins could suppress autophagy through this pathway as well. Perhaps this explains the cases in which detachment from the basement membrane actually promoted autophagy. However, it was shown that detachment of breast epithelial cells both increased autophagy and induced phosphorylation of eIF2α, a target of the ER kinase PERK and a protein capable of inducing autophagy [34].

AMPK activates autophagy by inhibiting TORC1 and activating Ulk1 via direct phosphorylation [49]. LKB1 localizes with engaged E-cadherin and activates AMPK signaling when ATP levels are low (Figure 3), providing a potential mechanism for cell-cell-interaction-mediated autophagy that might be relevant in detachment induced autophagy mechanisms in which cell-cell contact is maintained [50].

4. Conclusion

In many of the studies discussed above and in other diseases and models the effects of autophagy are very mitochondria-centric. For example, the Col6[-/-] mice phenotype was rescued by inhibiting mitochondrial depolarization, re-expressing Beclin or the pro-mitophagic protein Bnip3, or by starvation-induced autophagy [61,62]. Although not discussed in this chapter, it is interesting to note that Parkinson's disease is associated with mutations in several genes that target damaged mitochondria for mitophagy, and increasing evidence implicates mitophagy as having a role in retarding aging in general [75,76]. Considering that mitochondria can induce apoptosis and regulate the levels of ROS, one could imagine mitophagy as being a mechanism of survival during detachment or a method of maintaining cell integrity during adhesion.

Cell interaction with ECM and other cells of the same organism is imperative for - and essentially defines - multicellular organisms. Aside from being the scaffolding that organizes cells into functional macrostructures, the ECM extensively regulates cellular behavior [30]. The importance of these interactions is well demonstrated in the mounting evidence indicating that varied interactions with the ECM can actually differentially dictate the fate of stem cell differentiation [77]. Understanding the effects of the ECM on cell biology not only increases the understanding of normal organismal development and biology, but also diseases such as cancer or the various forms of muscular dystrophy [78,79]. With the current rise in understanding of autophagy in diseases and development, understanding how extracellular interactions and differentiation affect autophagy is an important avenue to explore in the future.

Nomenclature

3-MA - 3-methyladenine

AIF - apoptosis inducing factor

BAT - brown adipose tissue

Col6 - collagen VI gene

CFSE - 5,6-carboxyfluorescein diacetate succinimidyl ester

EB - embryoid body

ECM - extracellular matrix

LAMA2 - laminin α2 chain

Mdx - dystrophin gene

MEF - mouse embryo fibroblast

PrEC - prostate epithelial cell

SSC- squamous cell carcinomas

UCMD - Ulrich's Congenital Muscular Dystrophy

WAT- white adipose tissues

Acknowledgements

We wish to thank the Van Andel Research Institute (EAN, CKM) and the Van Andel Institute Graduate School (EAN) for their generous support. This work was also supported in part by an NIH/NCI grant CA154835 (CKM).

Author details

Eric A. Nollet[1,2] and Cindy K. Miranti[1*]

*Address all correspondence to: cindy.miranti@vai.org

1 Laboratory of Integrin Signaling and Tumorigensis, Van Andel Research Institute, Grand Rapids, Michigan, USA

2 Van Andel Institute Graduate School, Grand Rapids, Michigan, USA

References

[1] Gronowicz, G, Swift, H, & Steck, T. L. Maturation of the reticulocyte in vitro. J Cell Sci. (1984). http://www.ncbi.nlm.nih.gov/pubmed/6097593, 71, 177-97.

[2] Koury, M. J, Koury, S. T, Kopsombut, P, & Bondurant, M. C. In vitro maturation of nascent reticulocytes to erythrocytes. Blood. (2005). http://www.ncbi.nlm.nih.gov/pubmed/15528310, 105(5), 2168-74.

[3] Kundu, M, Lindsten, T, Yang, C. Y, Wu, J, Zhao, F, Zhang, J, et al. Ulk1 plays a critical role in the autophagic clearance of mitochondria and ribosomes during reticulocyte maturation. Blood. (2008). http://www.ncbi.nlm.nih.gov/pubmed/18539900, 112(4), 1493-502.

[4] Schweers, R. L, Zhang, J, Randall, M. S, Loyd, M. R, Li, W, Dorsey, F. C, et al. NIX is required for programmed mitochondrial clearance during reticulocyte maturation. Proc Natl Acad Sci U S A. (2007). http://www.ncbi.nlm.nih.gov/pubmed/18048346, 104(49), 19500-5.

[5] Ney, P. A. Normal and disordered reticulocyte maturation. Curr Opin Hematol. (2011). http://www.ncbi.nlm.nih.gov/pubmed/21423015, 18(3), 152-7.

[6] Zhang, J, Loyd, M. R, Randall, M. S, Waddell, M. B, Kriwacki, R. W, & Ney, P. A. A short linear motif in BNIP3L (NIX) mediates mitochondrial clearance in reticulocytes. Autophagy. (2012). http://www.ncbi.nlm.nih.gov/pubmed/22906961, 8(9), 1325-32.

[7] Zhang, J, Randall, M. S, Loyd, M. R, Dorsey, F. C, Kundu, M, Cleveland, J. L, et al. Mitochondrial clearance is regulated by Atg7-dependent and-independent mechanisms during reticulocyte maturation. Blood. (2009). http://www.ncbi.nlm.nih.gov/pubmed/19417210, 114(1), 157-64.

[8] Aerbajinai, W, Giattina, M, Lee, Y. T, Raffeld, M, & Miller, J. L. The proapoptotic factor Nix is coexpressed with Bcl-xL during terminal erythroid differentiation. Blood. (2003). http://www.ncbi.nlm.nih.gov/pubmed/12663450, 102(2), 712-7.

[9] Hanna, R. A, Quinsay, M. N, Orogo, A. M, Giang, K, Rikka, S, & Gustafsson, A. B. Microtubule-associated protein 1 light chain 3 (LC3) interacts with Bnip3 protein to selectively remove endoplasmic reticulum and mitochondria via autophagy. J Biol Chem. (2012). http://www.ncbi.nlm.nih.gov/pubmed/22505714, 287(23), 19094-104.

[10] Cebollero, E, Reggiori, F, & Kraft, C. Reticulophagy and ribophagy: regulated degradation of protein production factories. Int J Cell Biol. (2012). http://www.ncbi.nlm.nih.gov/pubmed/22481944

[11] Mortensen, M, Ferguson, D. J, Edelmann, M, Kessler, B, Morten, K. J, Komatsu, M, et al. Loss of autophagy in erythroid cells leads to defective removal of mitochondria and severe anemia in vivo. Proc Natl Acad Sci U S A. (2010). http://www.ncbi.nlm.nih.gov/pubmed/20080761, 107(2), 832-7.

[12] Pua, H. H, Guo, J, Komatsu, M, & He, Y. W. Autophagy is essential for mitochondrial clearance in mature T lymphocytes. J Immunol. (2009). http://www.ncbi.nlm.nih.gov/pubmed/19299702, 182(7), 4046-55.

[13] Stephenson, L. M, Miller, B. C, Ng, A, Eisenberg, J, Zhao, Z, Cadwell, K, et al. Identification of Atg5-dependent transcriptional changes and increases in mitochondrial mass in Atg5-deficient T lymphocytes. Autophagy. (2009). http://www.ncbi.nlm.nih.gov/pubmed/19276668, 5(5), 625-35.

[14] Dumitriu, I. E, Mohr, W, Kolowos, W, Kern, P, Kalden, J. R, & Herrmann, M. 5,6-carboxyfluorescein diacetate succinimidyl ester-labeled apoptotic and necrotic as well as detergent-treated cells can be traced in composite cell samples. Anal Biochem. (2001). http://www.ncbi.nlm.nih.gov/pubmed/11730350, 299(2), 247-52.

[15] Pua, H. H, Dzhagalov, I, Chuck, M, Mizushima, N, & He, Y. W. A critical role for the autophagy gene Atg5 in T cell survival and proliferation. The Journal of experimental medicine. (2007). http://www.ncbi.nlm.nih.gov/pubmed/17190837, 204(1), 25-31.

[16] Baerga, R, Zhang, Y, Chen, P. H, Goldman, S, & Jin, S. Targeted deletion of autophagy-related 5 (atg5) impairs adipogenesis in a cellular model and in mice. Autophagy. (2009). http://www.ncbi.nlm.nih.gov/pubmed/19844159, 5(8), 1118-30.

[17] Singh, R, Xiang, Y, Wang, Y, Baikati, K, Cuervo, A. M, Luu, Y. K, et al. Autophagy regulates adipose mass and differentiation in mice. The Journal of clinical investigation. (2009). http://www.ncbi.nlm.nih.gov/pubmed/19855132, 119(11), 3329-39.

[18] Zhang, Y, Goldman, S, Baerga, R, Zhao, Y, Komatsu, M, & Jin, S. Adipose-specific deletion of autophagy-related gene 7 (atg7) in mice reveals a role in adipogenesis. Proc Natl Acad Sci U S A. (2009). http://www.ncbi.nlm.nih.gov/pubmed/19910529, 106(47), 19860-5.

[19] Glick, D, Zhang, W, Beaton, M, Marsboom, G, Gruber, M, Simon, M. C, et al. BNip3 regulates mitochondrial function and lipid metabolism in the liver. Mol Cell Biol. (2012). http://www.ncbi.nlm.nih.gov/pubmed/22547685, 32(13), 2570-84.

[20] Lippens, S, Denecker, G, Ovaere, P, Vandenabeele, P, & Declercq, W. Death penalty for keratinocytes: apoptosis versus cornification. Cell Death Differ. (2005). Suppl http://www.ncbi.nlm.nih.gov/pubmed/16247497, 2, 1497-508.

[21] Gosselin, K, Deruy, E, Martien, S, Vercamer, C, Bouali, F, Dujardin, T, et al. Senescent keratinocytes die by autophagic programmed cell death. Am J Pathol. (2009). http://www.ncbi.nlm.nih.gov/pubmed/19147823, 174(2), 423-35.

[22] Aymard, E, Barruche, V, Naves, T, Bordes, S, Closs, B, Verdier, M, et al. Autophagy in human keratinocytes: an early step of the differentiation? Exp Dermatol. (2011). http://www.ncbi.nlm.nih.gov/pubmed/21166723, 20(3), 263-8.

[23] Klionsky, D. J, Abdalla, F. C, Abeliovich, H, Abraham, R. T, Acevedo-arozena, A, Adeli, K, et al. Guidelines for the use and interpretation of assays for monitoring autophagy. Autophagy. (2012). http://www.ncbi.nlm.nih.gov/pubmed/22966490, 8(4), 445-544.

[24] Xie, Z, & Bikle, D. D. The recruitment of phosphatidylinositol 3-kinase to the E-cadherin-catenin complex at the plasma membrane is required for calcium-induced phospholipase C-gamma1 activation and human keratinocyte differentiation. J Biol Chem. (2007). http://www.ncbi.nlm.nih.gov/pubmed/17242406, 282(12), 8695-703.

[25] Calautti, E, Li, J, Saoncella, S, Brissette, J. L, & Goetinck, P. F. Phosphoinositide 3-kinase signaling to Akt promotes keratinocyte differentiation versus death. J Biol Chem. (2005). http://www.ncbi.nlm.nih.gov/pubmed/16036919, 280(38), 32856-65.

[26] Kenny, P. A, Lee, G. Y, Myers, C. A, Neve, R. M, Semeiks, J. R, Spellman, P. T, et al. The morphologies of breast cancer cell lines in three-dimensional assays correlate with their profiles of gene expression. Molecular oncology. (2007). http://www.ncbi.nlm.nih.gov/pubmed/18516279, 1(1), 84-96.

[27] Fung, C, Lock, R, Gao, S, Salas, E, & Debnath, J. Induction of autophagy during extracellular matrix detachment promotes cell survival. Mol Biol Cell. (2008). http://www.ncbi.nlm.nih.gov/pubmed/18094039, 19(3), 797-806.

[28] Desbaillets, I, Ziegler, U, Groscurth, P, & Gassmann, M. Embryoid bodies: An in vitro model of mouse embryogenesis. Exp Physiol. (2000). http://www.ncbi.nlm.nih.gov/pubmed/11187960, 85(6), 645-51.

[29] Qi, Y, Tian, X, Liu, J, Han, Y, Graham, A. M, Simon, M. C, et al. Bnip3 and AIF coop-erate to induce apoptosis and cavitation during epithelial morphogenesis. J Cell Biol. (2012). http://www.ncbi.nlm.nih.gov/pubmed/22753893, 198(1), 103-14.

[30] Miranti, C. K, & Brugge, J. S. Sensing the environment: a historical perspective on in-tegrin signal transduction. Nature cell biology. (2002). Ehttp://www.ncbi.nlm.nih.gov/pubmed/11944041, 83-90.

[31] Sieg, D. J, Hauck, C. R, & Schlaepfer, D. D. Required role of focal adhesion kinase (FAK) for integrin-stimulated cell migration. J Cell Sci. (1999). Pt 16):2677-91. http://www.ncbi.nlm.nih.gov/pubmed/10413676

[32] Frisch, S. M, & Francis, H. Disruption of epithelial cell-matrix interactions induces apoptosis. J Cell Biol. (1994). http://www.ncbi.nlm.nih.gov/pubmed/8106557, 124(4), 619-26.

[33] Mizushima, N, & Levine, B. Autophagy in mammalian development and differentia-tion. Nature cell biology. (2010). http://www.ncbi.nlm.nih.gov/pubmed/20811354, 12(9), 823-30.

[34] Avivar-valderas, A, Salas, E, Bobrovnikova-marjon, E, Diehl, J. A, Nagi, C, Debnath, J, et al. PERK integrates autophagy and oxidative stress responses to promote surviv-al during extracellular matrix detachment. Mol Cell Biol. (2011). http://www.ncbi.nlm.nih.gov/pubmed/21709020, 31(17), 3616-29.

[35] Sandilands, E, Serrels, B, Mcewan, D. G, Morton, J. P, Macagno, J. P, Mcleod, K, et al. Autophagic targeting of Src promotes cancer cell survival following reduced FAK signalling. Nature cell biology. [10.1038/ncb2386]. (2012). http://www.ncbi.nlm.nih.gov/pubmed/22138575, 14(1), 51-60.

[36] Edick, M. J, Tesfay, L, Lamb, L. E, Knudsen, B. S, & Miranti, C. K. Inhibition of integ-rin-mediated crosstalk with epidermal growth factor receptor/Erk or Src signaling pathways in autophagic prostate epithelial cells induces caspase-independent death. Mol Biol Cell. (2007). http://www.ncbi.nlm.nih.gov/pubmed/17475774, 18(7), 2481-90.

[37] Lamb, L. E, Zarif, J. C, & Miranti, C. K. The androgen receptor induces integrin a6b1 to promote prostate tumor cell survival via NF-kB and Bcl-xL Independently of PI3K signaling. Cancer Res. (2011). http://www.ncbi.nlm.nih.gov/pubmed/21310825, 71(7), 2739-49.

[38] Friedland, J. C, Lakins, J. N, Kazanietz, M. G, Chernoff, J, Boettiger, D, & Weaver, V. M. a. a6b4 integrin activates Rac-dependent kinase 1 to drive NF-kB-dependent re-sistance to apoptosis in 3D mammary acini. J Cell Sci. (2007). Pt 20):3700-12. http://www.ncbi.nlm.nih.gov/pubmed/17911169, 21.

[39] Weaver, V. M, Lelievre, S, Lakins, J. N, Chrenek, M. A, Jones, J. C, Giancotti, F, et al. b4 integrin-dependent formation of polarized three-dimensional architecture confers

resistance to apoptosis in normal and malignant mammary epithelium. Cancer Cell. (2002). http://www.ncbi.nlm.nih.gov/pubmed/12242153, 2(3), 205-16.

[40] Zahir, N, Lakins, J. N, Russell, A, Ming, W, Chatterjee, C, Rozenberg, G. I, et al. Autocrine laminin-5 ligates a6b4 integrin and activates RAC and NF-kB to mediate anchorage-independent survival of mammary tumors. J Cell Biol. (2003). http://www.ncbi.nlm.nih.gov/pubmed/14691145, 163(6), 1397-407.

[41] Copetti, T, Bertoli, C, Dalla, E, Demarchi, F, & Schneider, C. p. RelA modulates BECN1 transcription and autophagy. Mol Cell Biol. (2009). http://www.ncbi.nlm.nih.gov/pubmed/19289499, 29(10), 2594-608.

[42] Zheng, Y. H, Tian, C, Meng, Y, Qin, Y. W, Du, Y. H, Du, J, et al. Osteopontin stimulates autophagy via integrin/CD44 and MAPK signaling pathways in vascular smooth muscle cells. J Cell Physiol. (2012). http://www.ncbi.nlm.nih.gov/pubmed/21374592, 38.

[43] Chen, G. C, Lee, J. Y, Tang, H. W, Debnath, J, Thomas, S. M, & Settleman, J. Genetic interactions between Drosophila melanogaster Atg1 and paxillin reveal a role for paxillin in autophagosome formation. Autophagy. (2008). http://www.ncbi.nlm.nih.gov/pubmed/17952025, 4(1), 37-45.

[44] Florey, O, Kim, S. E, Sandoval, C. P, Haynes, C. M, & Overholtzer, M. Autophagy machinery mediates macroendocytic processing and entotic cell death by targeting single membranes. Nature cell biology. (2011). http://www.ncbi.nlm.nih.gov/pubmed/22002674, 13(11), 1335-43.

[45] Overholtzer, M, Mailleux, A. A, Mouneimne, G, Normand, G, Schnitt, S. J, King, R. W, et al. A nonapoptotic cell death process, entosis, that occurs by cell-in-cell invasion. Cell. (2007). http://www.ncbi.nlm.nih.gov/pubmed/18045538, 131(5), 966-79.

[46] Lamb, L. E, & Knudsen, B. S. Miranti CK. E-cadherin-mediated survival of androgen-receptor-expressing secretory prostate epithelial cells derived from a stratified in vitro differentiation model. J Cell Sci. (2010). Pt 2):266-76. http://www.ncbi.nlm.nih.gov/pubmed/20048343

[47] Boussadia, O, Kutsch, S, Hierholzer, A, & Delmas, V. Kemler R. E-cadherin is a survival factor for the lactating mouse mammary gland. Mech Dev. (2002). http://www.ncbi.nlm.nih.gov/pubmed/12049767

[48] Li, L, Bennett, S. A, & Wang, L. Role of E-cadherin and other cell adhesion molecules in survival and differentiation of human pluripotent stem cells. Cell Adh Migr. (2012). http://www.ncbi.nlm.nih.gov/pubmed/22647941, 6(1), 59-70.

[49] Kim, J, Kundu, M, Viollet, B, & Guan, K. L. AMPK and mTOR regulate autophagy through direct phosphorylation of Ulk1. Nature cell biology. (2011). http://www.ncbi.nlm.nih.gov/pubmed/21258367, 13(2), 132-41.

[50] Sebbagh, M, Santoni, M. J, Hall, B, Borg, J. P, & Schwartz, M. A. Regulation of LKB1/ STRAD localization and function by E-cadherin. Current biology : CB. (2009). http:// www.ncbi.nlm.nih.gov/pubmed/19110428, 19(1), 37-42.

[51] Yang, P. M, Liu, Y. L, Lin, Y. C, Shun, C. T, Wu, M. S, & Chen, C. C. Inhibition of autophagy enhances anticancer effects of atorvastatin in digestive malignancies. Cancer Res. (2010). http://www.ncbi.nlm.nih.gov/pubmed/20876807, 70(19), 7699-709.

[52] Allamand, V, & Guicheney, P. Merosin-deficient congenital muscular dystrophy, autosomal recessive (MDC1A, MIM#156225, LAMA2 gene coding for alpha2 chain of laminin). Eur J Hum Genet. (2002). http://www.ncbi.nlm.nih.gov/pubmed/11938437, 10(2), 91-4.

[53] Miyagoe, Y, Hanaoka, K, Nonaka, I, Hayasaka, M, Nabeshima, Y, Arahata, K, et al. Laminin a2 chain-null mutant mice by targeted disruption of the Lama2 gene: a new model of merosin (laminin 2)-deficient congenital muscular dystrophy. FEBS Lett. (1997). http://www.ncbi.nlm.nih.gov/pubmed/9326364, 415(1), 33-9.

[54] Carmignac, V, Svensson, M, Korner, Z, Elowsson, L, Matsumura, C, Gawlik, K. I, et al. Autophagy is increased in laminin a2 chain-deficient muscle and its inhibition improves muscle morphology in a mouse model of MDC1A. Hum Mol Genet. (2011). http://www.ncbi.nlm.nih.gov/pubmed/21920942, 20(24), 4891-902.

[55] Chamberlain, J. S, Metzger, J, Reyes, M, Townsend, D, & Faulkner, J. A. Dystrophin-deficient mdx mice display a reduced life span and are susceptible to spontaneous rhabdomyosarcoma. FASEB J. (2007). http://www.ncbi.nlm.nih.gov/pubmed/ 17360850, 21(9), 2195-204.

[56] Pauly, M, Daussin, F, Burelle, Y, Li, T, Godin, R, Fauconnier, J, et al. AMPK activation stimulates autophagy and ameliorates muscular dystrophy in the mdx mouse diaphragm. Am J Pathol. (2012). http://www.ncbi.nlm.nih.gov/pubmed/22683340, 181(2), 583-92.

[57] Doe, J. A, Wuebbles, R. D, Allred, E. T, Rooney, J. E, Elorza, M, & Burkin, D. J. Transgenic overexpression of the a7 integrin reduces muscle pathology and improves viability in the dy(W) mouse model of merosin-deficient congenital muscular dystrophy type 1A. J Cell Sci. (2011). Pt 13):2287-97. http://www.ncbi.nlm.nih.gov/pubmed/ 21652631

[58] Sparks, S, Quijano-roy, S, Harper, A, Rutkowski, A, Gordon, E, Hoffman, E. P, et al. Congenital Muscular Dystrophy Overview. (1993). http://www.ncbi.nlm.nih.gov/ pubmed/20301468

[59] Irwin, W. A, Bergamin, N, Sabatelli, P, Reggiani, C, Megighian, A, Merlini, L, et al. Mitochondrial dysfunction and apoptosis in myopathic mice with collagen VI deficiency. Nature genetics. (2003). http://www.ncbi.nlm.nih.gov/pubmed/14625552, 35(4), 367-71.

[60] Merlini, L, Angelin, A, Tiepolo, T, Braghetta, P, Sabatelli, P, Zamparelli, A, et al. Cy-
      closporin A corrects mitochondrial dysfunction and muscle apoptosis in patients
      with collagen VI myopathies. Proc Natl Acad Sci U S A. (2008). http://
      www.ncbi.nlm.nih.gov/pubmed/18362356, 105(13), 5225-9.

[61] Palma, E, Tiepolo, T, Angelin, A, Sabatelli, P, Maraldi, N. M, Basso, E, et al. Genetic
      ablation of cyclophilin D rescues mitochondrial defects and prevents muscle apopto-
      sis in collagen VI myopathic mice. Hum Mol Genet. (2009). http://
      www.ncbi.nlm.nih.gov/pubmed/19293339, 18(11), 2024-31.

[62] Grumati, P, Coletto, L, Sandri, M, & Bonaldo, P. Autophagy induction rescues mus-
      cular dystrophy. Autophagy. (2011). http://www.ncbi.nlm.nih.gov/pubmed/
      21543891, 7(4), 426-8.

[63] Oak, S. A, Zhou, Y. W, & Jarrett, H. W. Skeletal muscle signaling pathway through
      the dystrophin glycoprotein complex and Rac1. J Biol Chem. (2003). http://
      www.ncbi.nlm.nih.gov/pubmed/12885773, 278(41), 39287-95.

[64] Wei, Y, Pattingre, S, Sinha, S, Bassik, M, & Levine, B. JNK1-mediated phosphoryla-
      tion of Bcl-2 regulates starvation-induced autophagy. Molecular cell. (2008). http://
      www.ncbi.nlm.nih.gov/pubmed/18570871, 30(6), 678-88.

[65] Yogev, O, Goldberg, R, Anzi, S, & Shaulian, E. Jun proteins are starvation-regulated
      inhibitors of autophagy. Cancer Res. (2010). http://www.ncbi.nlm.nih.gov/pubmed/
      20197466, 70(6), 2318-27.

[66] Li, D. D, Wang, L. L, Deng, R, Tang, J, Shen, Y, Guo, J. F, et al. The pivotal role of c-
      Jun NH2-terminal kinase-mediated Beclin 1 expression during anticancer agents-in-
      duced autophagy in cancer cells. Oncogene. (2009). http://www.ncbi.nlm.nih.gov/
      pubmed/19060920, 28(6), 886-98.

[67] Wei, Y, Sinha, S. C, & Levine, B. Dual Role of JNK1-mediated phosphorylation of
      Bcl-2 in autophagy and apoptosis regulation. Autophagy. (2008). http://
      www.ncbi.nlm.nih.gov/pubmed/18570871, 4(7), 949-51.

[68] Riol-blanco, L, Sanchez-sanchez, N, Torres, A, Tejedor, A, Narumiya, S, Corbi, A. L,
      et al. The chemokine receptor CCR7 activates in dendritic cells two signaling mod-
      ules that independently regulate chemotaxis and migratory speed. J Immunol. (2005).
      http://www.ncbi.nlm.nih.gov/pubmed/15778365, 174(7), 4070-80.

[69] Djavaheri-mergny, M, Amelotti, M, Mathieu, J, Besancon, F, Bauvy, C, & Codogno, P.
      Regulation of autophagy by NF-kB transcription factor and reactives oxygen species.
      Autophagy. (2007). http://www.ncbi.nlm.nih.gov/pubmed/17471012, 3(4), 390-2.

[70] Salminen, A, Hyttinen, J. M, Kauppinen, A, & Kaarniranta, K. Context-Dependent
      Regulation of Autophagy by IKK-NF-kB Signaling: Impact on the Aging Process. Int
      J Cell Biol. (2012). http://www.ncbi.nlm.nih.gov/pubmed/22899934

[71] Criollo, A, Senovilla, L, Authier, H, Maiuri, M. C, Morselli, E, Vitale, I, et al. The IKK complex contributes to the induction of autophagy. The EMBO journal. (2010). http:// www.ncbi.nlm.nih.gov/pubmed/19959994, 29(3), 619-31.

[72] Wu, Y. T, Tan, H. L, Huang, Q, Ong, C. N, & Shen, H. M. Activation of the PI3K-Akt-mTOR signaling pathway promotes necrotic cell death via suppression of autopha-gy. Autophagy. (2009). http://www.ncbi.nlm.nih.gov/pubmed/19556857, 5(6), 824-34.

[73] Chen, Y, & Klionsky, D. J. The regulation of autophagy- unanswered questions. J Cell Sci. (2011). Pt 2):161-70. http://www.ncbi.nlm.nih.gov/pubmed/21187343

[74] Mammucari, C, Schiaffino, S, & Sandri, M. Downstream of Akt: FoxO3 and mTOR in the regulation of autophagy in skeletal muscle. Autophagy. (2008). http:// www.ncbi.nlm.nih.gov/pubmed/18367868, 4(4), 524-6.

[75] Jin, S. M, & Youle, R. J. PINK1- and Parkin-mediated mitophagy at a glance. J Cell Sci. (2012). Pt 4):795-9. http://www.ncbi.nlm.nih.gov/pubmed/22448035

[76] Seo, A. Y, Joseph, A. M, Dutta, D, Hwang, J. C, Aris, J. P, & Leeuwenburgh, C. New insights into the role of mitochondria in aging: mitochondrial dynamics and more. J Cell Sci. (2010). Pt 15):2533-42. http://www.ncbi.nlm.nih.gov/pubmed/20940129

[77] Guilak, F, Cohen, D. M, Estes, B. T, Gimble, J. M, Liedtke, W, & Chen, C. S. Control of stem cell fate by physical interactions with the extracellular matrix. Cell Stem Cell. (2009). http://www.ncbi.nlm.nih.gov/pubmed/19570510, 5(1), 17-26.

[78] Lu, P, Weaver, V. M, & Werb, Z. The extracellular matrix: a dynamic niche in cancer progression. J Cell Biol. (2012). http://www.ncbi.nlm.nih.gov/pubmed/22351925, 196(4), 395-406.

[79] Davies, K. E, & Nowak, K. J. Molecular mechanisms of muscular dystrophies: old and new players. Nat Rev Mol Cell Biol. (2006). http://www.ncbi.nlm.nih.gov/ pubmed/16971897, 7(10), 762-73.

Chapter 10

Time Flies: Autophagy During Ageing in *Drosophila*

Sebastian Wolfgang Schultz, Andreas Brech and
Ioannis P. Nezis

Additional information is available at the end of the chapter

1. Introduction

1.1. Ageing

The process of ageing compromises the age-associated decrease in fertility, gradual loss of function, and increased vulnerability to disease, which progressively diminishes the capability of an organism to survive [1-3]. Unsurprisingly, in the past years it has been of great interest to understand which factors influence this inevitable and complex process. As a result a wide array of molecular and cellular damages has been identified and shown to accumulate during ageing. The lifelong accumulation of such damages will eventually result in frailty and disease [4]. The variety of identified age-dependent damages has given rise to different theories for molecular ageing mechanisms. These mechanisms include decreased cellular capacity to deal with DNA damage, and decline in cellular division capacity, which is linked to the progressive shortening of telomeres upon each cell cycle. Also an increased accumulation of damaged mitochondria and the involved increase in reactive oxygen species (ROS) production and decline in ATP synthesis has been shown to occur over time (reviewed in [5]). One of the phenotypic hallmarks of aged cells is the intracellular accumulation of damaged proteins and therefore protein turnover/protein degradation has attracted attention over the last years [2].

At the same time, forward genetics have allowed to investigate single gene alterations and their influence on lifespan of whole organisms. Even though the ageing process is without doubt influenced by stochastic and environmental factors, single gene mutations were shown to extend lifespan in worms, flies, and mice, suggesting the existence of a central process of ageing [6, 7]. Many of the genetic manipulations that alter longevity affect metabolism, nutrient sensing and stress response pathways. As all these pathways are connected to autophagy (an important player also in protein turnover), the question about the role of autophagy in ageing has come more and more to the fore. In this chapter we will focus on how research conducted in the excellent genetic model system *Drosophila melanogaster* has contributed to understand more about the interplay of autophagy and ageing.

2. Autophagy

Autophagy, which literally means "self-eating" (coined by Nobel Laureate Christian de Duve in 1963), allows cells to digest cytosolic components via lysosomal degradation. Autophagy and the Ubiquitin Proteasome System (UPS) constitute together the main cellular pathways for protein and organelle turnover [8, 9]. Today, three different classes of autophagy are distinguished: microautophagy, chaperone-mediated autophagy (CMA), and macroautophagy.

During microautophagy, which is mainly studied in yeast (containing vacuoles instead of lysosomes), cytoplasmic material is delivered to the vacuolar lumen by direct invagination of the vacuolar boundary membrane and budding of autophagic bodies into the vacuolar lumen [10]. The molecular mechanisms underlying microautophagy in eukaryotic cells are largely unknown. However, Cuervo and colleagues described a microautophagy-like process (named endosomal microautophagy, e-MI) in mammalian cells, whereby soluble cytosolic proteins are selectively taken up by late endosomes/multivesicular bodies (MVBs). The cargo selection in e-MI depends on the chaperone Hsc70 and electrostatic interactions with the endosomal membrane [11]. Hsc70 is also involved in chaperone-mediated autophagy (CMA), in which cytosolic cargo is selectively recognized, bound by the lysosome-associated membrane type protein 2A (LAMP-2A) and finally taken up by the lysosome, thereby allowing for direct lysosomal degradation of cytosolic proteins. The requirement of protein unfolding and the binding of LAMP2-A is characteristic for CMA and thereby distinguishes CMA from e-MI [11, 12]. So far, CMA has not been investigated in *Drosophila melanogaster*. The third common type of autophagy, macroautophagy (henceforth referred to as autophagy), is highly conserved from yeast to mammalian cells [8]. Autophagy allows for cytosolic bulk degradation of long-lived macromolecules and organelles. Morphologically this process was already described in the 1960s but it was not before several decades later when genetic screens in *Saccharomyces cerevisiae* identified multiple genes involved in autophagy and thereby allowed to investigate the molecular mechanisms in further detail [13, 14]. Genetic screens in *S. cerevisiae* have since then led to the identification of numerous autophagy-related (ATG) genes and many homologs have been identified and characterized in higher eukaryotes [15]. In general, autophagy can be divided into three steps: 1) induction/nucleation; 2) expansion; and 3) maturation [16].

The formation of a cytosolic double membrane structure called the phagophore (also called isolation membrane) is an important step of autophagy initiation. It is subject of discussion about the origin of this initial autophagic membrane. Independent experiments identified ER, Golgi, or the outer membrane of mitochondria to contribute to the phagophore double membrane [17, 18]. Cytosolic components are enwrapped during the growth of the phagophore. Closure of the phagophore completes this engulfment and gives rise to a new structure called the autophagosome. These newly formed autophagosomes will further mature and subsequently fuse with lysosomes where the captured cytosolic constituents will be degraded. Autophagy can achieve several purposes; it scavenges the cytosol from macromolecules and organelles but also provides a way to supply the cells with amino acids and if necessary with energy once the recycled amino acids are converted into intermediates of the tricarboxylic acid cycle (TCA) [15, 18-20]. It is therefore of little surprise that the autophagic machinery, which under normal conditions is running on low basal levels, can be set in motion by several intra- and extracellular stress factors, such as starvation, ER-stress, hypoxia and pathogen invasion [15]. Besides non-selective cytosolic bulk-degradation, autophagy is also implicated in selective turnover in yeast, a pathway known as the cytoplasm-to-vacuole targeting (CVT) pathway [21]. In analogy, cargo selective degradation of aggregated proteins (aggrephagy [22]), mitochondria (mitophagy [23]), ribosomes (ribophagy [24]), peroxisomes (pexophagy [25]), endoplasmic reticulum (reticulophagy [26]) and many more have been reported for mammalian systems [27]. The role of selective autophagy in ageing will be further addressed in a separate section of this chapter.

Several protein complexes are involved along the path from initiation to completion of autophagy. Induction of autophagy in *Drosophila* requires the Ser/Thr kinase Atg1 that forms a complex with Atg13. Phosphorylation of Atg13 by Atg1 directs phagophore initiation through a complex containing the class III PI(3)-kinase Vps34, the Ser/Thr kinase Vps15 and Atg6 (Beclin1 in mammals). The activation of this complex leads to localized generation of phosphatidylinositol-3-phosphate (PI3P), a critical step in autophagy. In mammalian systems, this core complex has several known interaction partners, e.g. Atg14, Ambra1, UVRAG, or Rubicon, that are all involved in autophagy. Several of these mammalian genes have ortho-logues in *Drosophila*, however their involvement in autophagy remains to be shown (reviewed in [28]). UVRAG has recently been found to be important in the regulation of Notch levels in the context of organ rotation during development. This role of UVRAG is coupled to endocytic degradation of Notch and, in this context, not to autophagy [29]. Autophagosome formation requires the ubiquitin-like proteins Atg12 and Atg8 and their respective ubiquitin-like conjugation systems [30]. Atg8 is processed by the cysteine protease Atg4 and covalently linked to phosphatidylethanolamine (PE) through the action of the E1 activating enzyme Atg7, the E2 activating enzyme Atg3 and the E3 like Atg12-Atg5-Atg16 complex, which is found at the phagophore membrane. The E3 like Atg12-Atg5-Atg16 complex itself requires also Atg7 and the E2 activating enzyme Atg10 for its assembly (reviewed in [31]). Once Atg8 is activated and lipid-conjugated it is localized to both sides of the phagophore and Atg4 later only removes the portion residing at the cytosolic side prior to autophagosome-lysosome/endosome fusion. It has also been reported that Atg8 can modulate the size of autophagosomes by influencing membrane curvature. For all these reasons, activation of Atg8/LC3 is widely used to monitor autophagy [15, 32]. The process from autophagy initiation until autolysosome formation is schematically illustrated in figure 1.

It is believed that stepwise fusion of autophagosomes with different endosomal populations account for maturation and culminates in the fusion with lysosomes, the organelle responsi-ble for degradation [33]. Such stepwise fusion is supported by the findings that impairment of ESCRT machinery results in reduced autolysosome formation, measured as decrease in lysotracker staining and accumulation of Atg8 positive punctate respectively [34, 35]. Similar accumulation of autophagosomes can be seen in flies with mutant *Drosophila* deep orange (dor) and dvps16A [36, 37]. Both proteins are known to play important roles in endocytic trafficking.

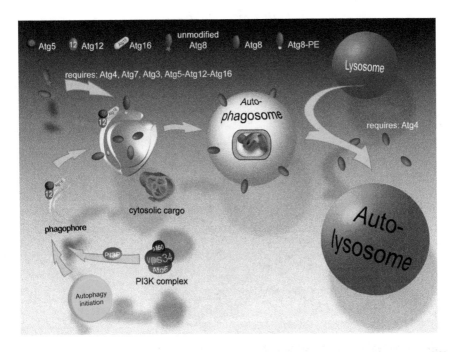

Figure 1. Schematic illustration of macroautophagy. Upon autophagy initiation the PI3K complex generates PI3P, which is then provided at high local concentrations at the initial step of phagophore membrane formation. The ubiquitin like proteins Atg12 and Atg8 with their respective conjugation system are recruited and activated once the phagophore is formed. Membrane expansion leads to phagophore maturation, which is finalized by vesicle closure and thereby autophagosome formation. This vesicle can fuse with different endocytic compartments or directly with lysosomes, forming autolysosomes. There, phagophore-sequestered cytosolic cargo is degraded and macromolecules can be recycled back to the cytosol. For further details see section 2 and references therein.

3. Role of autophagy in development, homeostasis and ageing

In general, the role of autophagy is predominantly described as cytoprotective. Intensive research over the last decade has increased our understanding of multiple cellular events that involve autophagy, e.g. dealing with low nutrient levels, development and morphogenesis, response to oxidative stress, turnover of protein aggregates and damaged organelles, immune response and lately also cell signaling (figure 2). Altogether the picture has emerged that the role and regulation of autophagy is extremely dependent on the cellular context [20, 38, 39].

With this review we therefore want to highlight what is known from research conducted in *Drosophila*, a model organism, which allows for elegant genetic manipulations of the cellular setup in a multi cellular organism.

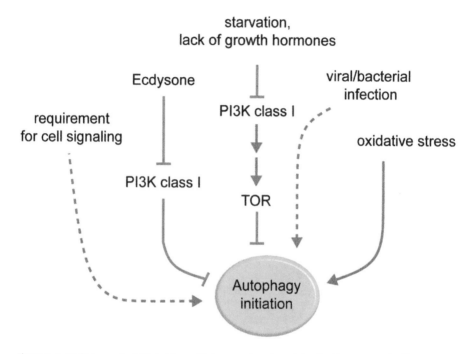

Figure 2. Autophagy can be initiated by multiple ways. Autophagy is involved in a variety of different cellular events (e.g. development, survival under conditions of low nutrient levels, oxidative stress response, immune response, and cell signaling), which requires several ways to initiate the core autophagy machinery (dashed lines: the exact pathway is still uncertain, however autophagy is shown to be upregulated as downstream effect). For further details see section 3 and references therein.

In *Drosophila*, autophagy plays a pivotal role during development and is crucial for a wide range of developmental processes. Cell growth depends on nutrients provided by autophagy as seen in the fat body. On the other hand autophagy has been reported to be necessary for targeted cell death and removal of tissue, e.g. during oogenesis and development of gut salivary glands.

More than ten years ago, both the class I phosphatidylinositol 3-kinase (PI3K) and the serine/threonine kinase Target of Rapamycin (TOR) have been shown to control a signaling network that is important for development (reviewed in [40]). The growth of cells and tissues does require energy and building blocks. Hormones, such as insulin have been identified as important signals in order to meet these requirements by e.g. upregulation of protein synthesis. Already in 2003, Tom Neufeld speculated about the role of catabolic processes, such as autophagy, to be important in development. This idea was supported by previous findings that established a connection between reduced basal autophagic protein turnover and cellular growth as well as that Apg6p, the yeast homologue of the tumor suppressor gene *Beclin 1*, is required for autophagy in yeast (reviewed in [40]). Furthermore, it was already shown that insulin, as well as class I PI3Ks can, besides their effect on protein synthesis, inhibit autophagic protein turnover, providing a plausible molecular link between autophagy and cell growth [41, 42]. Therewith the stage was set for two important findings published in 2004, revealing the regulation of programmed autophagy in the fat body and the importance for functional autophagy in cell growth [43, 44]. The levels of the hormone 20-hydroxyecdysone (ecdysone) rise during the development of *Drosophila*, leading to inactivation of the class I PI3K and subsequent autophagy activation [43]. This initiation of autophagy is necessary in order to supply the developing *Drosophila* larva/pupa with nutrients and to maintain survival and growth. The protective role of autophagy in this context is dominant over its otherwise known role in growth suppression [44, 45]. Noteworthy, dor (Deep orange), the *Drosophila* homolog to Vps18, can influence autophagy in the fat body in two separate ways. Dor is necessary for secretion of ecdysone from the salivary glands, thereby influencing the levels of this hormone. However, dor is also important in the fusion of autophagosomes with lysosomes, thereby directly controlling autophagy [36]. Autophagy in the fat body is dependent on the PI3K Vps34 [34]. Vps34 was initially identified to be involved in vacuolar protein sorting (Vps) in yeast [46]. Flies lacking Vps34 or its regulatory subunit, the protein kinase Vps15 (also referred to as p150), are hampered in their ability to initiate autophagy upon starvation in the fat body and die during development [34, 47]. Interestingly, the absence of Atg7 does not lead to lethality in the developing fly. Atg7 deficient flies have severe defects in autophagy but nevertheless are viable. However, such flies are short lived, show signs of accelerated ageing in the form of ubiquitin-positive aggregates in degenerating neurons and have very low resistance to nutrient deprivation and oxidative stress. This underscores the necessity of functional autophagy for cellular homeostasis and stress survival in the adult fly [48].

A very different aspect of autophagy during development has been revealed in the context of programmed cell death. Autophagy is upregulated during the reorganisation of the salivary gland and gut [49, 50]. Inhibition of autophagy in salivary glands by activating the class I PI3K pathway reduces salivary gland cell degradation. In contrast, induction of autophagy in salivary gland cells results in premature cell death and it was shown that this cell death is dependent on both caspases and autophagy [49]. Similar events can be seen in the midgut. Even though caspases are highly expressed, the canonical apoptotis pathway is not required for midgut removal. Inhibition of autophagy on the other hand, impairs midgut degradation and simultaneously decreases caspase activity [50]. Additional ways how cell death and autophagy are connected are pointed out by the findings that autophagy can selectively degrade survival factors and thereby initiate cell death. During late oogenesis, autophagy is necessary to degrade the apoptosis inhibitor dBruce in nurse cells. Nurse cells lack the, under normal conditions typical, fragmentation of DNA and caspase-3 activity in the absence of autophagy [51]. A similar principle for cell death control is suggested by the finding that the valosin-containing protein (vcp), a ubiquitin-selective AAA chaperone, is required for degradation of the apoptosis inhibitor DIAP1 during regulated degeneration of dendrites of class IV dendritic arborisation neurons [52]. It was already shown before that vcp is necessary for autophagy [53]. Altogether, this implies a role for autophagy in activating apoptosis by selective degradation of apoptosis inhibitors. It will be interesting to see if such a mechanism is limited to the programmed reorganization events during development or if this is a strategy employed even in other cellular contexts. If this is a general mechanism to initiate cell death, autophagic degradation of apoptosis inhibitors might become an interesting strategy for developing drugs aimed for cancer treatment.

The role of autophagy in *Drosophila* is not limited to development but instead autophagy is also important for various aspects during lifetime of eclosed flies. Any organism needs to be able to cope with oxidative stress, which itself is tightly linked to ageing [5]. In *Drosophila*, Jun N-terminal kinase (JNK) can protect the gut from oxidative toxicity due to feeding on paraquat, a well-established oxidative stress inducer. In addition, genetic upregulation of the JNK pathway extends lifespan of flies in a Foxo dependent manner [54, 55]. This cell protective effect of JNK is mediated by the transcriptional activation of autophagy. JNK cannot protect flies from oxidative toxicity when Atg1 or Atg6 activity is reduced [56].

A different putative way for autophagy to protect cells from oxidative stress is given by its involvement in the selective degradation of damaged mitochondria, termed mitophagy. So far, mitophagy has not been directly shown to occur in *Drosophila*, nevertheless several findings indicate that mitophagy also happens in flies. Studies in *Drosophila* have suggested that the E3 ubiquitin ligase Parkin normally facilitates mitochondrial fission and/or inhibits fusion [57]. In addition the PTEN-induced putative kinase protein 1 (PINK1) has been shown to genetically interact with Parkin in flies, and results from experiments in *Drosophila* S2 cells revealed that PINK1 is required for the recruitment of Parkin to damaged mitochondria leading to their degradation [58]. Interestingly, the finding that the level of the protrusion factor mitofusin (mfn) increased in the absence of PINK1 or Parkin, suggests that mfn might be ubiquitinated by Parkin, which can serve as putative label and targeting signal for degradation of damaged mitochondria [58]. In yeast and mammals it has been shown that ubiquitination of mitochondrial proteins by Parkin results in autophagic degradation of mitochondria (reviewed in [23]). This role of Parkin in the removal of damaged mitochondria might also explain the muscle degeneration, mitochondrial pathology and reduced lifespan in *parkin* mutant flies [59]. Lately, it has been reported that mitochondrial protein misfolding in *Drosophila* leads to degradation of mitochondria and that accumulation of an unfolded protein in the mitochondria phenocopies flies with mutations in PINK1 and Parkin. The requirement of Ref(2)P (refractory to Sigma P, the *Drosophila* homolog of p62) for this mitochondrial turnover resembles mitophagy as described in mammalian systems [60]. However, it remains to be proven that the turnover of damaged mitochondria in flies really is conducted by autophagy, hence that mitophagy also occurs in *Drosophila*.

Without doubt autophagy is crucial for cellular homeostasis and it is therefore of no surprise that autophagy is also induced upon viral or bacterial infections as both lead to changes in the intracellular environment. Flies with impaired autophagy are hampered in their immune defence. Even though this role of autophagy is much more studied in mammalian system, there are 4 different reports that highlight an involvement of autophagy in the *Drosophila* immune response. When autophagy was impaired by the expression of RNAi against Atg5, Atg7, or Atg12, *Drosophila* displays a decreased resistance to injected *Escherichia coli*, which manifests in higher titers of *E. coli* and reduced survival rates. Interestingly, knockdown of any of these three Atg genes did not shorten lifespan of uninjected flies [61]. The latter finding is not in line with findings from Atg7 deficient flies, which show a significant shortening of life span [48]. Even though the conditional knockdown of Atg7 did lead to a decrease in lysotracker staining, a sign for reduced autophagy, it cannot be excluded that some remaining Atg7 activity is enough in order to allow for basal autophagy and thereby not altering lifespan. It can be expected that such basal autophagy is more severely affected in flies completely missing the gene for Atg7. On the other hand there is also the possibility that Atg7$^{-/-}$ flies already accumulate cell damage during development that might allow them to hatch normally but still will give them a severe survival disadvantage right from the start.

Autophagy does not only protect against bacterial but also against viral infection as shown in the case of the mammalian viral pathogen vesicular stomatitis virus (VSV) [62]. Autophagy protects flies against VSV by decreasing viral replication. Repression of autophagy has the contrary effect, increased viral replication and pathogenesis. The authors of this study were able to pinpoint the PI3K/Akt pathway to be responsible for autophagy regulation upon VSV infection [62]. Flies infected with *Mycobacterium marinum* are dependent on autophagy in order for mycobacteria drug treatment to be successful. *Drosophila* lacking the gene for Atg7 had a reduced survival rate upon *Mycobacterium marinum* infection and this phenotype could not be rescued with the help of antimycobacterial treatment [63].

An additional involvement of autophagy in immunity was found in the cortical remodelling of hemocytes (*Drosophila* blood cells). Integrin-mediated hemocyte spreading and Rho1-induced cell protrusions require continuous autophagy. As a consequence, flies with impaired autophagy in their hemocytes show severe defects in recruiting hemocytes to epidermal wounds. Furthermore, this study identified Ref(2)P to be crucial for functional autophagy, which suggests selective autophagy (see below) to be involved in this process [64]. The requirement for selective autophagic turnover of single proteins to maintain cellular homeostasis has been implicated in several different cellular contexts. E.g. activated rhodopsin is degraded via the endosomal pathway and mutations in rhodopsin leading to hampered endocytic turnover results in retinal degeneration [65]. Autophagy has also been connected to the turnover of activated rhodopsin and mutations in Atg7 or Atg8, or genes necessary for proper autophagosome formation, result in light-dependent retinal degeneration [66].

Another example for the necessity of functional selective autophagic degradation of proteins for proper homeostasis is given in muscle tissue maintenance. There, chaperone-assisted selective autophagy is necessary to remove contraction-induced damaged filamin from Z-discs in order to prevent Z disk disintegration and progressive muscle weakness in flies [67].

Autophagy also serves several functions in neuron plasticity and homeostasis. An interesting finding was that synapse development is controlled by autophagy via the E3 ubiquitin ligase highwire. Highwire inhibits neuromuscular junction growth and is itself a substrate for selective autophagic turnover, indicating that autophagy activity might lead to synaptic overgrowth [68]. Tian et al. identified Rae1 to bind to highwire and thereby protecting highwire from autophagic degradation [69]. The link between autophagy and synaptic growth at the neuromuscular junction is further strengthened by the observation that ROS can act as signaling molecules and mediate synaptic growth. At the same time, high ROS levels activate the JNK pathway, a previously reported activator of autophagy. As impairment of autophagy results in decreased synaptic size in a *Drosophila* model, whereas activation of autophagy has the opposite effect, one can speculate that ROS mediated synaptic growth is mediated by activation of JNK and subsequent autophagy (reviewed in [70]).

Off course, autophagy also plays a major role in simply keeping the cells "clean" by enabling the cells to turn over the cytosol. This recycling effect is especially pronounced in post-mitotic cells. Flies that are mutant for Atg8a are severely hampered in their efficiency to eliminate cellular material, which can be observed as an increase in ubiquitinated proteins and the increased presence of electron dense protein aggregates in young fly brains when investigated with transmission electron microscopy. Moreover, such Atg8a mutant flies display a drastic decrease of lifespan [71]. In addition, a very recent report by Fouillet et al., reveals an autophagy-mediated decrease of apoptosis in neurons upon mild ER-stress and further underscore a cytoprotective role of autophagy, that potentially can prolong survival of the whole organism [72]. Taken together, these data outline the versatile role of autophagy in homeostasis and normal survival of flies.

4. Autophagy and neurodegeneration

Ageing is a major risk factor for the development of neurodegenerative diseases and over the last decade autophagy has been implicated in many neurodegenerative diseases, such as Huntington's disease (HD), Parkinson's disease (PD), amyotrophic lateral sclerosis (ALS), or Alzheimer's disease (AD). Many neurodegenerative diseases share the common phenotype of accumulations of protein aggregates [73]. Before reviewing the role of autophagy in *Drosophila* models for neurodegeneration we first want to give a short overview of key findings on the link between autophagy and neurodegeneration as known from mammalian systems and patient data.

Both, HD and PD are connected to elevated autophagy. In case of HD, autophagy can only be triggered by a mutant form of huntingtin that is prone to aggregate but not by wildtype huntingtin. Cytosolic aggregates of α-synuclein, the protein involved in PD, can be degraded by macroautophagy and CMA [74-77]. In ALS loss of motor neurons deprives patients of voluntary controlled muscle movements. The disease is associated with ubiquitinated, p62 positive protein inclusions of TDP-43 (TAR DNA binding protein 43) or SOD1 (superoxide dismutase 1) or rare mutations in a subunit of the ESCRT complex [78, 79]. A defective ESCRT complex in its turn has been shown to result in autophagosome accumulation [80], but also point mutations of the p150 subunit of dynactin resulting in defects in the transport machinery along microtubules have been implicated in ALS. Transport along microtubules is necessary for autophagosome-lysosome fusion and therefore crucial for functional autophagy [81, 82]. Extensive alterations in macroautophagy can also be found in patients with AD. An immuno-electron microscopy study on neocortical biopsies from AD patients identified autophago-somes, multivesicular bodies, multilamellar bodies, and cathepsin-containing autophagolysosomes as the predominant organelles that occupied most of the cytosol of dystrophic neurites. Autophagy was detected in cell bodies with neurofibrillary pathology and associated with a relative depletion of mitochondria and other organelles. The authors of this study speculated that the accumulation of immature autophagic vacuoles results from impaired transport to and fusion with lysosomes thereby hampering the protective effects of autophagy [83]. Disruption of lysosomal proteolysis in primary mouse cortical neurons by inhibiting cathepsins, or by supressing lysosomal acidification, impairs transport of autolyso-somes, endosomes and lysosomes, and leads to accumulation of these structures within dystrophic axonal swellings. Such a phenotype can also be seen in numerous mouse models of AD. The phenotype is not caused by general disruption of the axonal transport machinery, as mitochondria and cathepsin-lacking organelles were not influenced in their movements. Axonal dystrophy is reversed once lysosomal function is restored [84].

In the past, several independent groups have established *Drosophila* models for neurodege-nerative diseases and/or investigated the role of aggregating proteins implied in neurodege-nerative diseases in flies. Remarkably, already in 1982 Stark and Carlson characterized the degenerative phenotypes evoked by a mutant form of the rdgB (retinal-degeneration-B) protein in the fly compound eye and found amongst others lysosome-like bodies and vacuoles suggesting involvement of autophagy [85]. The compound eye of flies displays a highly structured order and degenerative properties of protein aggregates can easily be monitored as impairments of this structure. Expression of mutant huntingtin containing a polyQ-expansion of 120 glutamine leads to degeneration of the eye. However, treatment with rapamycin, an activator of autophagy, reduces this phenotype [86]. Treatment with new small-molecule enhancers (SMER) of the cytosolic effects of rapamycin, which were shown to induce autophagy in mammalian cells, also protected flies from polyQ huntingtin induced neurode-generation [87]. Instead of treating flies with rapamycin in order to inhibit TOR by pharma-cological means, Wang et al. highlighted the importance of TOR in neurodegeneration by genetical manipulations. Hyperactivation of TOR, achieved by expression of the TOR kinase activator Ras homologue enriched in brain protein (Rheb) or introduction of mutations in the TOR inhibitor dTsc1 increased age- and light-dependent photoreceptor loss [88]. The authors of this study were able to exclude TORs effects on growth to be responsible for this photore-ceptor degeneration but instead pointed out autophagy as the downstream signaling of TOR mediating photoreceptor cell death. Activation of autophagy by overexpressing Atg1 protect-ed not only cells from age- and light dependent photoreceptor degeneration, but also photo-receptor cells which either produced 120 polyQ-huntingtin or lacked a functional *Drosophila phospholipase C* gene *norpA* respectively. Both latter manipulations are commonly used to model neurodegeneration [88]

Macroautophagy in flies can also be upregulated by Rab5 over-expression and this approach also mitigates polyQ-huntingtin mediated degeneration in the eye [89]. However, there is a fine line between beneficial and detrimental consequences of autophagy activation in the context of neurodegeneration as shown in a dentatorubralpallidoluysian atrophy (DRPLA) fly model [90]. This model is built upon the expression of atrophin with a polyQ expansion and is characterized by lysosomal dysfunction and blocked autophagosome-lysosome fusion, hence reduced autophagic flux [90]. Even though introduction of a mutant form of Atg1 intensified the neurodegenerative phenotype, upregulation of autophagy in this system had no rescuing effect but, in some case, even had the opposite outcome and increased neurodegeneration [90]. In other words, autophagy plays an important role in scavenging polyQ atrophin from the cytosol, but is only of beneficial nature as long as autophagy can proceed all the way to lysosomal degradation. Reaching a rate-limiting step in the autophagy cycle can have negative effects on the outcome of autophagy initiation. Work from the same lab also identified a mechanism how polyQ atrophin itself impairs autophagic flux. PolyQ-atrophin inhibits the tumor suppressor fat, which under normal conditions protects from neurodegeneration through the Hippo kinase cascade and subsequent increases autophagy [91]. How Hippo exactly activates autophagy is not completely understood yet [92]. Data obtained from studies in the salivary gland suggests that the phosphorylation of Warts (wts), a substrate of the Hippo kinase, acts upstream of TOR and thereby regulates autophagy [93]. It also has been reported that the Hippo pathway can directly interact with LC3 (the mammalian homolog of Atg8) and thereby initiate autophagy [92].

An additional, interesting link between polyQ sequence derived neurodegeneration and macroautophagy is given by puromycin-sensitive aminopeptidase (PSA). PSA is the only cytosolic enzyme capable of digesting polyQ sequences and it is therefore not surprising that there is inverse correlation between PSA expression and severity of neurodegeneration, e.g. over-expression of PSA has protective effects in cells expressing polyQ expanded ataxin-3, mutant α-synuclein and mutant superoxide dismutase (SOD) [94]. It comes as a surprise though that this beneficial role of PSA is mediated by its activation of macroautophagy rather than its role in degrading polyQ aggregates and thereby making them available for proteasomal degradation, although the putative involvement of the proteasome in cell protection in this process remains to be further understood [94].

A different way to induce neurodegeneration is to inhibit proteasomal function. Interestingly, proteasome impairment can be compensated for by autophagy, a rescue that depends on the histone deacetylase 6 (HDAC6) [95]. A protective role of autophagy in context of neurodegeneration was also demonstrated in a genetic screen conducted in *Drosophila* with pathogenic Ataxin-3-induced neurodegeneration. Knockdown of Atg5 in these flies reverts the polyQ containing Ataxin-3 mediated toxicity. Testing the effects of identified neurodegeneration-suppressors on autophagy revealed that these factors had different impact on autophagy. The authors of this study proposed a model in which some neurodegeneration-suppressors induce autophagy, thereby contributing to protein clearance whereas others mitigate autophagy in order to counteract autophagic cell death [96]. The role of autophagy in removal of protein aggregates in neurodegenerative diseases was further confirmed by the finding that depletion of subunits of the ESCRT complex in flies intensifies the toxic effects exerted by polyQ-expanded huntingtin [97]. Depletion of ESCRT subunits has autophagy inhibition as consequence, which manifests in accumulation of protein aggregates containing ubiquitinated proteins, p62 and Alfy [98].

The Alzheimer's disease related peptide $A\beta_{1-42}$ also induces neurodegeneration, mediated by age-dependent autophagy-lysosomal injury in a *Drosophila* model of AD [99]. The age dependence was shown to be of high importance as brain ageing is accompanied by an increasingly defective autophagy-lysosomal system and accumulation of dysfunctional autophagosomes and autolysosomes. As a consequence intracellular membranes and organelles are damaged. The expression of $A\beta_{1-42}$ resulted in similar changes already in young *Drosophila* and this raised the question if chronic deterioration of the autophagy-lysosomal system by $A\beta_{1-42}$ simply accelerates brain ageing [100]. This concept is supported the finding that expression of autophagy genes decreases with age, and disruption of the autophagy pathway reduces lifespan of flies [71].

5. Autophagy and its role in lifetime extension

The rate of ageing is reciprocally linked to lifespan and therefore are interventions that extend longevity of an organism the most direct indication that ageing is slowed down [101]. One well established, and long known intervention that extends lifespan is dietary restriction (DR), the limitation of food intake below the *ad libidum* level without malnutrition. DR has successfully been proven to extend lifespan in every organism tested, including yeast, worms, flies and rodents. In addition, DR not only extends lifespan, even the occurrence of age-associated pathologies, e.g. cardiovascular disease, multiple kinds of cancer, neurodegeneration, are drastically reduced or at least postponed in animal models [102]. The possibility to perform forward genetics in different model organisms has boosted the general understanding of underlying molecular mechanisms how DR, and other life extending interventions, can execute their effects. Studies in *Caenorhabditis elegans* by Cynthia Kenyon and co-workers have already almost two decades ago showed how mutations in the single gene daf-2 (the insulin receptor homologue in *C. elegans*) can increase survival by more than two-fold and that such extended survival is dependent on a second gene, namely daf-16 (a forkhead transcription factor) [103, 104]. Since then the role of nutrient-sensing pathways in ageing has been addressed by many independent groups, which has helped to identify numerous proteins that are crucial in lifespan determination. Amongst other pathways, both the insulin/insulin-like growth factor (IGF) and the Target of Rapamycin (TOR) network have been shown to be important modulators of longevity (reviewed in [101, 105, 106]). The fact that both these networks also are involved in the regulation of autophagy emphasizes a putative role of autophagy in lifetime extension and has been addressed in *Drosophila* by several groups.

Simonsen and co-workers showed that downregulation of autophagy genes in *Drosophila* neural tissue is part of the normal ageing process. This is accompanied by accumulation of insoluble ubiquitinated proteins (IUPs). Impairment of autophagy due to mutations in Atg8a aggravates the occurrence of IUPs at earlier time points and lowers survival rates [71]. As lipid-conjugation of Atg8 is essential for nucleation and phagophore elongation it can be speculated that Atg8 is a limiting factor in autophagic turnover. The over-expression of Atg8 in the central nervous system of *Drosophila* indeed extends average and maximum life span by approx. 50% [71]. Flies not only live longer upon Atg8a over-expression, but also showed a higher tolerance to oxidative stress and lower occurrence of IUPs [71]. Interestingly, the longevity promoting effect of Atg8a over-expression cannot be seen when over-expression is initiated during development but decreases over time as seen in flies where Atg8a expression was driven by the early pan-neural driver line *Elav-Gal4* [71]

The question if IUPs are cause or a consequence of the ageing process remains to be answered though. Albeit, the age-dependent accumulation of ubiquitinated proteins that are positive for Ref(2)P, a protein necessary for cargo recognition in selective autophagy, can be employed as conserved marker of neuronal ageing and progressive autophagic defects [107].

Also Atg7 was recently reported to extend life span when over-expressed in neuronal tissues of flies [108]. The life-extending effect of Atg7 is not as pronounced when compared to Atg8. This might be due to different capabilities in inducing autophagic turnover, or non-autophagy related side effects of either Atg7 or Atg8a.

Proteostasis is not only important in neuronal tissues but also in muscles of flies. With increasing age polyubiquitinated proteins accumulate that co-localise with Ref(2)P in muscles and the cumulative appearance of such aggregates has been demonstrated to impair muscle fitness [109]. The build-up of such aggregates can be reverted in muscles by the constitutive activation of the transcription factor FOXO and its target 4E-BP (eukaryotic translation initiation factor 4E binding protein). Interestingly, the activation of FOXO/4E-BP signaling in muscles is sufficient to extend lifespan of the whole organism [109]. Furthermore it has been shown that the FOXO/4E-BP dependent delay in protein aggregate accumulation in muscles depends on functional autophagy, suggesting promotion of basal autophagy upon FOXO/4E-BP signaling [109]. The autophagy dependent beneficial effect of FOXO is well in line with earlier findings that revealed FOXO to be capable to upregulate autophagy [110]. In addition, the translational repressor 4E-BP is known to be upregulated upon DR and to mediate enhanced mitochondrial function and life span extension in *Drosophila* [111]. As already mentioned earlier, autophagy has a known role in the selective turnover of damaged mito-chondria in yeast and mammals, and it is therefore tempting to speculate that autophagy can promote longevity by improving mitochondrial function in a FOXO/4E-BP dependent manner, however this remains to be proven.

Ageing in *Drosophila* can also be manipulated by pharmacological means. Feeding the TOR inhibitory drug rapamycin, a well-described drug for human use, significantly increases lifespan and resistance to starvation as well as the oxidative stress inducer paraquat [112]. Rapamycin fails to extend the lifespan of flies with downregulated Atg5 suggesting that autophagy has to be active in order for rapamycin to slow down ageing [112]. The finding that inhibition of TOR increases lifespan in *Drosophila* is well in line with earlier studies demon-strating that mutant, inactive TOR or over-expression of the TOR inhibitors dTsc1 or dTsc2 extend longevity [113, 114]. However, the specific role of autophagy was not addressed in those two studies.

Keeping *Drosophila* on food supplemented with the polyamine spermidine promotes increased longevity and this effect has been shown to be autophagy dependent, since depletion of Atg7 abrogates this anti-ageing effect [115].

Taken together, all these data indicate an anti-ageing effect of autophagy, however caution is advised in trying to merely upregulate autophagy pharmacologically in order to counter-act ageing. Autophagy is essential for the recycling of cellular content, which can serve two general purposes: autophagy can unburden cells from hazards by removal of those and autophagy can provide cells with new building blocks for cellular survival. During the lifetime of an organism, autophagy will most certainly switch forth and back between those roles. In order to completely understand the complex role of autophagy in ageing it is therefore important to understand the regulation and cellular outcome of autophagy in a tissue and time dependent manner.

6. Selective autophagy and ageing

In the following section we want to shed some light on the current knowledge about the selective removal of cellular contents by autophagy in *Drosophila melanogaster*. Above, we have already discussed some examples of selective autophagy in normal ageing and homeostasis. We therefore will focus more on the mechanistic insights of selective autophagy and what is known so far about the role of selective autophagy explicitly in ageing of *Drosophila*.

Selective autophagy in the form of CVT has been known in yeast for a long time and has gained major attention in mammalian systems over the last years. Selectivity requires crucial, additional steps to the above described autophagy process: cargo has to be recognized by specific receptors and must be delivered to the autophagic machinery.

Ubiquitin has emerged as a molecule to tag proteins that are determined for degradation [116]. Conjugation of ubiquitin depends on a complex reaction cascade that requires activation of ubiquitin (by E1 enzymes), conjugation (E2 ubiquitin conjugating enzyme), and ligation of ubiquitin with a target substrate (E3 ubiquitin ligase). As a result, ubiquitin is covalently bound via an isopeptide bond between the C-terminal glycine of ubiquitin and the ε-amino group of a lysine residue on the substrate protein. Substrate specificity is given by the E3 ubiquitin ligase that specifically recognizes a protein substrate and brings it to the E2 ubiquitin conjugating enzyme. A wide spectrum of E1, E2, and E3 enzymes provide cells with selectivity for this signaling machinery [117]. Ubiquitin itself contains seven lysine residues enabling ubiquitin to self-attach, thereby forming a polyubiquitin tag. The best-characterised linkages occur via K48, targeting the substrate for proteasomal degradation, and via K63, which is preferred by ubiquitin-binding autophagy receptors. Furthermore, K63 ubiquitination has been reported to be a potent enhancer of inclusion formation and leads to substrate degradation via the autophagy/lysosome degradation pathway [116, 118-120]. Also more atypical sites for polyubiquitination, such as K6 or K29, have been reported but the exact role of these ubiquitin chains is still poorly understood [121].

Taken together, ubiquitin conjugation offers several possibilities to flag proteins and organelles in different ways by variation of chain length and various sites for ubiquitin self-attachment and thereby act as a signal for distinct subsequent cellular processing. Molecular links between ubiquitinated proteins and autophagy were identified in form of the cargo receptors sequestosome marker SQSTM1/p62 and NBR1 (neighbour of BRCA1 gene) [122]. The conserved functional homologue for p62/NBR1 in *Drosophila* is Ref(2)P. Ref(2)P is a 599 amino acid long protein with an N-terminal Phox Bem1p (PB1) domain, followed by a ZZ-type Zinc finger domain and a C-terminal UBA (ubiquitin-associated) domain [123]. The PB1 domain allows for self- and hetero-oligomerisation, while the UBA domain enables Ref(2)P to recognize and directly interact with ubiquitin. Both domains are necessary for formation of protein aggregates normally found in brains of adult *Drosophila* [124]. Flies mutant for Atg8 display an increased amount of deposited protein aggregates in the brain, however such aggregates are absent in double-mutant Atg8/Ref(2)P flies [124]. This suggests that Ref(2)P is a selective cargo receptor for selective autophagy in *Drosophila*, similar as its homologue p62 in mammals. This is supported by the presence of a putative LIR (LC3 interacting) domain in Ref(2)P as identified by bioinformatics analysis [122]. The LIR domain is known to be essential for p62 to interact with LC3, but it remains to be elucidated if Ref(2)P really interacts with Atg8 via its putative LIR domain. Independent of the absence of final proof of direct interaction between Atg8 and Ref(2)P, protein aggregations containing Ref(2)P serve as excellent markers for neuronal ageing and autophagic defects in *Drosophila* [107].

Filimonenko et al. were able to identify the mammalian phosphatidylinositol-3-phosphate (PI3P) binding protein Alfy (PI3P-binding Autophagy-linked FYVE domain protein) to be actively involved in autophagic degradation of polyglutamine (polyQ) expanded, aggregated proteins [125]. Albeit harbouring a FYVE domain Alfy is usually not found on endosomes but instead resides in the nucleus decorating the nuclear membrane. The presence of ubiquitinated, aggregated proteins in the cytosol leads to relocalization of Alfy to these aggregates [126]. Alfy can directly interact with p62 and Atg5 [125, 127]. *In vitro*, Alfy is necessary to recruit Atg5 to polyQ protein aggregates. In addition, Alfy scaffolds the Atg5-Atg12-Atg16L complex to p62- and ubiquitin-positive polyQ inclusions [125]. The Atg5-Atg12-Atg16L complex on the other hand is important for LC3 lipidation [128]. Taken together, all these interactions allow for LC3 lipidation in close spatial proximity to ubiquitinated, aggregated proteins and explain the absence of other cytosolic components in aggregate filled autophagosomes [125]. Primary neurons expressing polyQ Htt (Huntingtin) have fewer polyQ inclusions upon ectopic Alfy expression. These results were confirmed *in vivo* with a *Drosophila* model where polyQ production provokes a phenotype that is due to toxicity. The outcome of polyQ-mediated toxicity was much milder once bchs (blue cheese, the *Drosophila* homologue of Alfy) was co-expressed [125]. Reduced levels of bchs in mutant flies had opposite effects and led to shortened live span and extensive neurodegeneration [129]. It remains to be elucidated if Alfy/bchs directly recognizes ubiquitinated aggregates or if this interaction is mediated by p62/Ref(2)P [22].

Accumulation of damaged mitochondria and increased production of ROS are generally believed to account for age associated pathologies [5]. The efficiency of selective removal of damaged mitochondria, mitophagy, might therefore play a major role in the outcome of the ageing process. Although several lines of evidence suggest the existence of mitophagy in *Drosophila* (see section 3) the molecular details in flies still have to be further unravelled.

7. Summary and outlook

The cytoprotective role of autophagy has been shown in many different cellular contexts and induction of autophagy by either pharmacological or genetical means has life extending effects. However, research conducted in *Drosophila* has also identified situations during development when autophagy is necessary for controlled tissue removal and cell death initiation. These two rather contrary roles, cytoprotection versus cell death initiation, highlight the complexity of the autophagy pathway and also underscore the importance to understand the molecular mechanisms by which autophagy exerts its role. As autophagy most likely is regulated in a tissue and time dependent manner it is of great interest to pinpoint those time points and tissues in which autophagy has the biggest impact on the general ageing processes.

Ageing is not only influenced by one single pathway but in contrary is a multifaceted process. Age is a major risk factor for a variety of diseases, e.g. neurodegenerative diseases, metabolic syndrome, cancer and more. In the past, extensive research has been undertaken to model neurodegenerative diseases in the fruitfly and has helped to push our understanding, not the least concerning the involvement of autophagy, to new levels. Today, *Drosophila* is getting growing attention as cancer model and it will be exciting to follow future research in order to get new insights from *Drosophila melanogaster* about the complex role of autophagy in cancer. By putting several different pieces of puzzle together, *Drosophila* already has helped us to get a clearer picture about the role of autophagy in various aspects of ageing and for sure the fruitfly will continue to help the research community to reveal more of this complex picture in the future.

Author details

Sebastian Wolfgang Schultz[1,2], Andreas Brech[1,2] and Ioannis P. Nezis[3]

*Address all correspondence to: I.Nezis@warwick.ac.uk

1 Department of Biochemistry, Institute for Cancer Research, Oslo University Hospital, The Norwegian Radium Hospital, Oslo, Norway

2 Centre for Cancer Biomedicine, Faculty of Medicine, University of Oslo, Oslo, Norway

3 School of Life Sciences, University of Warwick, Coventry, United Kingdom

References

[1] Kirkwood, T. B, & Austad, S. N. Why do we age? Nature, (2000). , 233-238.

[2]   Vellai, T. Autophagy genes and ageing. Cell death and differentiation, (2009). , 94-102.

[3]   Lionaki, E, Markaki, M, & Tavernarakis, N. Autophagy and ageing: Insights from invertebrate model organisms. Ageing research reviews, (2012).

[4]   Kirkwood, T. B. A systematic look at an old problem. Nature, (2008). , 644-647.

[5]   Kirkwood, T. B. Understanding the odd science of aging. Cell, (2005). , 437-447.

[6]   Guarente, L, & Kenyon, C. Genetic pathways that regulate ageing in model organisms. Nature, (2000). , 255-262.

[7]   Hekimi, S, & Guarente, L. Genetics and the specificity of the aging process. Science, (2003). , 1351-1354.

[8]   Reggiori, F, & Klionsky, D. J. Autophagy in the eukaryotic cell. Eukaryotic cell, (2002). , 11-21.

[9]   De Duve, C. The lysosome. Scientific American, (1963). , 64-72.

[10] Uttenweiler, A, et al. The vacuolar transporter chaperone (VTC) complex is required for microautophagy. Molecular biology of the cell, (2007). , 166-175.

[11] Sahu, R, et al. Microautophagy of cytosolic proteins by late endosomes. Developmental cell, (2011). , 131-139.

[12] Cuervo, A. M. Chaperone-mediated autophagy: selectivity pays off. Trends in endocrinology and metabolism: TEM, (2010). , 142-150.

[13] Tsukada, M, & Ohsumi, Y. Isolation and characterization of autophagy-defective mutants of Saccharomyces cerevisiae. FEBS letters, (1993). , 169-174.

[14] Thumm, M, et al. Isolation of autophagocytosis mutants of Saccharomyces cerevisiae. FEBS letters, (1994). , 275-280.

[15] He, C, & Klionsky, D. J. Regulation mechanisms and signaling pathways of autophagy. Annual review of genetics, (2009). , 67-93.

[16] Yamamoto, A, & Simonsen, A. The elimination of accumulated and aggregated proteins: A role for aggrephagy in neurodegeneration. Neurobiol Dis, (2010).

[17] Mcewan, D. G, & Dikic, I. Not all autophagy membranes are created equal. Cell, (2010). , 564-566.

[18] Tooze, S. A, & Yoshimori, T. The origin of the autophagosomal membrane. Nature cell biology, (2010). , 831-835.

[19] Seglen, P. O, Gordon, P. B, & Holen, I. Non-selective autophagy. Seminars in cell biology, (1990). , 441-448.

[20] Mizushima, N, & Komatsu, M. Autophagy: renovation of cells and tissues. Cell, (2011). , 728-741.

[21] Yorimitsu, T, & Klionsky, D. J. Autophagy: molecular machinery for self-eating. Cell death and differentiation, (2005). Suppl 2: , 1542-1552.

[22] Knaevelsrud, H, & Simonsen, A. Fighting disease by selective autophagy of aggregate-prone proteins. FEBS letters, (2010). , 2635-2645.

[23] Youle, R. J, & Narendra, D. P. Mechanisms of mitophagy. Nature reviews. Molecular cell biology, (2011). , 9-14.

[24] Kraft, C, et al. Mature ribosomes are selectively degraded upon starvation by an autophagy pathway requiring the Ubp3ubiquitin protease. Nature cell biology, (2008). p. 602-10., Bre5p.

[25] Dunn, W. A, et al. Pexophagy: the selective autophagy of peroxisomes. Autophagy, (2005). , 75-83.

[26] Bernales, S, Mcdonald, K. L, & Walter, P. Autophagy counterbalances endoplasmic reticulum expansion during the unfolded protein response. PLoS biology, (2006). , e423.

[27] Klionsky, D. J, et al. How shall I eat thee? Autophagy, (2007). , 413-416.

[28] Chang, Y. Y, & Neufeld, T. P. Autophagy takes flight in Drosophila. FEBS letters, (2010). , 1342-1349.

[29] Lee, G, et al. UVRAG is required for organ rotation by regulating Notch endocytosis in Drosophila. Developmental biology, (2011). , 588-597.

[30] Ohsumi, Y. Molecular dissection of autophagy: two ubiquitin-like systems. Nature reviews. Molecular cell biology, (2001). , 211-216.

[31] Mcphee, C. K, & Baehrecke, E. H. Autophagy in Drosophila melanogaster. Biochimica et biophysica acta, (2009). , 1452-1460.

[32] Simonsen, A, & Tooze, S. A. Coordination of membrane events during autophagy by multiple class III PI3-kinase complexes. The Journal of cell biology, (2009). , 773-782.

[33] Razi, M, Chan, E. Y, & Tooze, S. A. Early endosomes and endosomal coatomer are required for autophagy. The Journal of cell biology, (2009). , 305-321.

[34] Juhasz, G, et al. The class III PI(3)K Vps34 promotes autophagy and endocytosis but not TOR signaling in Drosophila. The Journal of cell biology, (2008). , 655-666.

[35] Rusten, T. E, et al. ESCRTs and Fab1 regulate distinct steps of autophagy. Current biology : CB, (2007). , 1817-1825.

[36] Lindmo, K, et al. A dual function for Deep orange in programmed autophagy in the Drosophila melanogaster fat body. Experimental cell research, (2006). , 2018-2027.

[37] Pulipparacharuvil, S, et al. Drosophila Vps16A is required for trafficking to lyso-
      somes and biogenesis of pigment granules. Journal of cell science, (2005). Pt 16): ,
      3663-3673.

[38] Backues, S. K, & Klionsky, D. J. Autophagy gets in on the regulatory act. Journal of
      molecular cell biology, (2011). , 76-77.

[39] Rubinsztein, D. C, Marino, G, & Kroemer, G. Autophagy and aging. Cell, (2011). ,
      682-695.

[40] Neufeld, T. P. Body building: regulation of shape and size by PI3K/TOR signaling
      during development. Mechanisms of development, (2003). , 1283-1296.

[41] Petiot, A, et al. Distinct classes of phosphatidylinositol 3'-kinases are involved in sig-
      naling pathways that control macroautophagy in HT-29 cells. The Journal of biologi-
      cal chemistry, (2000). , 992-998.

[42] Pfeifer, U. Inhibition by insulin of the formation of autophagic vacuoles in rat liver. A
      morphometric approach to the kinetics of intracellular degradation by autophagy.
      The Journal of cell biology, (1978). , 152-167.

[43] Rusten, T. E, et al. Programmed autophagy in the Drosophila fat body is induced by
      ecdysone through regulation of the PI3K pathway. Developmental cell, (2004). ,
      179-192.

[44] Scott, R. C, Schuldiner, O, & Neufeld, T. P. Role and regulation of starvation-induced
      autophagy in the Drosophila fat body. Developmental cell, (2004). , 167-178.

[45] Lee, S. B, et al. ATG1, an autophagy regulator, inhibits cell growth by negatively reg-
      ulating S6 kinase. EMBO reports, (2007). , 360-365.

[46] Banta, L. M, et al. Organelle assembly in yeast: characterization of yeast mutants de-
      fective in vacuolar biogenesis and protein sorting. The Journal of cell biology,
      (1988). , 1369-1383.

[47] Lindmo, K, et al. The PI 3-kinase regulator Vps15 is required for autophagic clear-
      ance of protein aggregates. Autophagy, (2008). , 500-506.

[48] Juhasz, G, et al. Atg7-dependent autophagy promotes neuronal health, stress toler-
      ance, and longevity but is dispensable for metamorphosis in Drosophila. Genes & de-
      velopment, (2007). , 3061-3066.

[49] Berry, D. L, & Baehrecke, E. H. Growth arrest and autophagy are required for saliva-
      ry gland cell degradation in Drosophila. Cell, (2007). , 1137-1148.

[50] Denton, D, et al. Autophagy, not apoptosis, is essential for midgut cell death in Dro-
      sophila. Current biology : CB, (2009). , 1741-1746.

[51] Nezis, I. P, et al. Autophagic degradation of dBruce controls DNA fragmentation in nurse cells during late Drosophila melanogaster oogenesis. The Journal of cell biology, (2010). , 523-531.

[52] Rumpf, S, et al. Neuronal remodeling and apoptosis require VCP-dependent degradation of the apoptosis inhibitor DIAP1. Development, (2011). , 1153-1160.

[53] Ju, J. S, et al. Valosin-containing protein (VCP) is required for autophagy and is disrupted in VCP disease. The Journal of cell biology, (2009). , 875-888.

[54] Wang, M. C, Bohmann, D, & Jasper, H. JNK signaling confers tolerance to oxidative stress and extends lifespan in Drosophila. Developmental cell, (2003). , 811-816.

[55] Wang, M. C, Bohmann, D, & Jasper, H. JNK extends life span and limits growth by antagonizing cellular and organism-wide responses to insulin signaling. Cell, (2005). , 115-125.

[56] Wu, H, Wang, M. C, & Bohmann, D. JNK protects Drosophila from oxidative stress by trancriptionally activating autophagy. Mechanisms of development, (2009). , 624-637.

[57] Deng, H, et al. The Parkinson's disease genes pink1 and parkin promote mitochondrial fission and/or inhibit fusion in Drosophila. Proceedings of the National Academy of Sciences of the United States of America, (2008). , 14503-14508.

[58] Ziviani, E, Tao, R. N, & Whitworth, A. J. Drosophila parkin requires PINK1 for mitochondrial translocation and ubiquitinates mitofusin. Proceedings of the National Academy of Sciences of the United States of America, (2010). , 5018-5023.

[59] Greene, J. C, et al. Mitochondrial pathology and apoptotic muscle degeneration in Drosophila parkin mutants. Proceedings of the National Academy of Sciences of the United States of America, (2003). , 4078-4083.

[60] Pimenta de Castro I., et al., Genetic analysis of mitochondrial protein misfolding in Drosophila melanogaster. Cell death and differentiation, (2012). , 1308-1316.

[61] Ren, C, Finkel, S. E, & Tower, J. Conditional inhibition of autophagy genes in adult Drosophila impairs immunity without compromising longevity. Experimental gerontology, (2009). , 228-235.

[62] Shelly, S, et al. Autophagy is an essential component of Drosophila immunity against vesicular stomatitis virus. Immunity, (2009). , 588-598.

[63] Kim, J. J, et al. Host cell autophagy activated by antibiotics is required for their effective antimycobacterial drug action. Cell host & microbe, (2012). , 457-468.

[64] Kadandale, P, et al. Conserved role for autophagy in Rho1-mediated cortical remodeling and blood cell recruitment. Proceedings of the National Academy of Sciences of the United States of America, (2010). , 10502-10507.

[65]   Chinchore, Y, Mitra, A, & Dolph, P. J. Accumulation of rhodopsin in late endosomes triggers photoreceptor cell degeneration. PLoS genetics, (2009). , e1000377.

[66]   Midorikawa, R, et al. Autophagy-dependent rhodopsin degradation prevents retinal degeneration in Drosophila. The Journal of neuroscience : the official journal of the Society for Neuroscience, (2010). , 10703-10719.

[67]   Arndt, V, et al. Chaperone-assisted selective autophagy is essential for muscle main-tenance. Current biology : CB, (2010). , 143-148.

[68]   Shen, W, & Ganetzky, B. Autophagy promotes synapse development in Drosophila. The Journal of cell biology, (2009). , 71-79.

[69]   Tian, X, et al. Drosophila Rae1 controls the abundance of the ubiquitin ligase High-wire in post-mitotic neurons. Nature neuroscience, (2011). , 1267-1275.

[70]   Milton, V. J, & Sweeney, S. T. Oxidative stress in synapse development and function. Developmental neurobiology, (2012). , 100-110.

[71]   Simonsen, A, et al. Promoting basal levels of autophagy in the nervous system en-hances longevity and oxidant resistance in adult Drosophila. Autophagy, (2008). , 176-184.

[72]   Fouillet, A, et al. ER stress inhibits neuronal death by promoting autophagy. Autoph-agy, (2012). , 915-926.

[73]   Knaevelsrud, H, & Simonsen, A. Fighting disease by selective autophagy of aggre-gate-prone proteins. FEBS Lett, (2010). , 2635-2645.

[74]   Nagata, E, et al. Autophagosome-like vacuole formation in Huntington's disease lymphoblasts. Neuroreport, (2004). , 1325-1328.

[75]   Ravikumar, B, et al. Inhibition of mTOR induces autophagy and reduces toxicity of polyglutamine expansions in fly and mouse models of Huntington disease. Nat Gen-et, (2004). , 585-595.

[76]   Webb, J. L, et al. Alpha-Synuclein is degraded by both autophagy and the protea-some. J Biol Chem, (2003). , 25009-25013.

[77]   Cuervo, A. M, et al. Impaired degradation of mutant alpha-synuclein by chaperone-mediated autophagy. Science, (2004). , 1292-1295.

[78]   Parkinson, N, et al. ALS phenotypes with mutations in CHMP2B (charged multive-sicular body protein 2B). Neurology, (2006). , 1074-1077.

[79]   Talbot, K, & Ansorge, O. Recent advances in the genetics of amyotrophic lateral scle-rosis and frontotemporal dementia: common pathways in neurodegenerative dis-ease. Human molecular genetics, (2006). Spec (2), R182-R187.

[80]   Rusten, T. E, & Stenmark, H. How do ESCRT proteins control autophagy? Journal of cell science, (2009). Pt 13): , 2179-2183.

[81] Munch, C, et al. Point mutations of the subunit of dynactin (DCTN1) gene in ALS. Neurology, (2004). p. 724-6., 150.

[82] Rubinsztein, D. C. The roles of intracellular protein-degradation pathways in neuro-degeneration. Nature, (2006). , 780-786.

[83] Nixon, R. A, et al. Extensive involvement of autophagy in Alzheimer disease: an immuno-electron microscopy study. Journal of neuropathology and experimental neurology, (2005). , 113-122.

[84] Lee, S, Sato, Y, & Nixon, R. A. Lysosomal proteolysis inhibition selectively disrupts axonal transport of degradative organelles and causes an Alzheimer's-like axonal dystrophy. The Journal of neuroscience : the official journal of the Society for Neuroscience, (2011). , 7817-7830.

[85] Stark, W. S, & Carlson, S. D. Ultrastructural pathology of the compound eye and optic neuropiles of the retinal degeneration mutant (w rdg BKS222) Drosophila melanogaster. Cell and tissue research, (1982). , 11-22.

[86] Ravikumar, B, et al. Inhibition of mTOR induces autophagy and reduces toxicity of polyglutamine expansions in fly and mouse models of Huntington disease. Nature genetics, (2004). , 585-595.

[87] Sarkar, S, et al. Small molecules enhance autophagy and reduce toxicity in Huntington's disease models. Nature chemical biology, (2007). , 331-338.

[88] Wang, T, Lao, U, & Edgar, B. A. TOR-mediated autophagy regulates cell death in Drosophila neurodegenerative disease. The Journal of cell biology, (2009). , 703-711.

[89] Ravikumar, B, et al. Rab5 modulates aggregation and toxicity of mutant huntingtin through macroautophagy in cell and fly models of Huntington disease. Journal of cell science, (2008). Pt 10): , 1649-1660.

[90] Nisoli, I, et al. Neurodegeneration by polyglutamine Atrophin is not rescued by induction of autophagy. Cell death and differentiation, (2010). , 1577-1587.

[91] Napoletano, F, et al. Polyglutamine Atrophin provokes neurodegeneration in Drosophila by repressing fat. The EMBO journal, (2011). , 945-958.

[92] Calamita, P, & Fanto, M. Slimming down fat makes neuropathic hippo: the Fat/Hippo tumor suppressor pathway protects adult neurons through regulation of autophagy. Autophagy, (2011). , 907-909.

[93] Dutta, S, & Baehrecke, E. H. Warts is required for PI3K-regulated growth arrest, autophagy, and autophagic cell death in Drosophila. Current biology : CB, (2008). , 1466-1475.

[94] Menzies, F. M, et al. Puromycin-sensitive aminopeptidase protects against aggregation-prone proteins via autophagy. Human molecular genetics, (2010). , 4573-4586.

[95]   Pandey, U. B, et al. HDAC6 rescues neurodegeneration and provides an essential link between autophagy and the UPS. Nature, (2007). , 859-863.

[96]   Bilen, J, & Bonini, N. M. Genome-wide screen for modifiers of ataxin-3 neurodegeneration in Drosophila. PLoS genetics, (2007). , 1950-1964.

[97]   Rusten, T. E, et al. ESCRTing autophagic clearance of aggregating proteins. Autophagy, (2007).

[98]   Filimonenko, M, et al. Functional multivesicular bodies are required for autophagic clearance of protein aggregates associated with neurodegenerative disease. The Journal of cell biology, (2007). , 485-500.

[99]   Ling, D, et al. Abeta42-induced neurodegeneration via an age-dependent autophagic-lysosomal injury in Drosophila. PloS one, (2009). , e4201.

[100] Ling, D, & Salvaterra, P. M. Brain aging and Abeta$_{1-42}$ neurotoxicity converge via deterioration in autophagy-lysosomal system: a conditional Drosophila model linking Alzheimer's neurodegeneration with aging. Acta neuropathologica, (2011). , 183-191.

[101] Partridge, L, et al. Ageing in Drosophila: the role of the insulin/Igf and TOR signaling network. Experimental gerontology, (2011). , 376-381.

[102] Bishop, N. A, & Guarente, L. Genetic links between diet and lifespan: shared mechanisms from yeast to humans. Nature reviews. Genetics, (2007). , 835-844.

[103] Kenyon, C, et al. A C. elegans mutant that lives twice as long as wild type. Nature, (1993). , 461-464.

[104] Lin, K, et al. daf-16: An HNF-3/forkhead family member that can function to double the life-span of Caenorhabditis elegans. Science, (1997). , 1319-1322.

[105] Kapahi, P, et al. With TOR, less is more: a key role for the conserved nutrient-sensing TOR pathway in aging. Cell metabolism, (2010). , 453-465.

[106] Kenyon, C. J. The genetics of ageing. Nature, (2010). , 504-512.

[107] Bartlett, B. J, et al. Ref(2)P and ubiquitinated proteins are conserved markers of neuronal aging, aggregate formation and progressive autophagic defects. Autophagy, (2011). p. 572-83., 62.

[108] Chen, S. F, et al. Autophagy-related gene 7 is downstream of heat shock protein 27 in the regulation of eye morphology, polyglutamine toxicity, and lifespan in Drosophila. Journal of biomedical science, (2012). , 52.

[109] Demontis, F, & Perrimon, N. FOXO/4E-BP signaling in Drosophila muscles regulates organism-wide proteostasis during aging. Cell, (2010). , 813-825.

[110] Juhasz, G, et al. Gene expression profiling identifies FKBP39 as an inhibitor of autophagy in larval Drosophila fat body. Cell death and differentiation, (2007). , 1181-1190.

[111] Zid, B. M, et al. E-BP extends lifespan upon dietary restriction by enhancing mitochondrial activity in Drosophila. Cell, (2009). , 149-160.

[112] Bjedov, I, et al. Mechanisms of life span extension by rapamycin in the fruit fly Drosophila melanogaster. Cell metabolism, (2010). , 35-46.

[113] Kapahi, P, et al. Regulation of lifespan in Drosophila by modulation of genes in the TOR signaling pathway. Current biology : CB, (2004). , 885-890.

[114] Luong, N, et al. Activated FOXO-mediated insulin resistance is blocked by reduction of TOR activity. Cell metabolism, (2006). , 133-142.

[115] Eisenberg, T, et al. Induction of autophagy by spermidine promotes longevity. Nature cell biology, (2009). , 1305-1314.

[116] Kirkin, V, et al. A role for ubiquitin in selective autophagy. Molecular cell, (2009). , 259-269.

[117] Hershko, A, & Ciechanover, A. The ubiquitin system. Annual review of biochemistry, (1998). , 425-479.

[118] Tan, J. M, et al. Lysine 63-linked ubiquitination promotes the formation and autophagic clearance of protein inclusions associated with neurodegenerative diseases. Human molecular genetics, (2008). , 431-439.

[119] Welchman, R. L, Gordon, C, & Mayer, R. J. Ubiquitin and ubiquitin-like proteins as multifunctional signals. Nature reviews. Molecular cell biology, (2005). , 599-609.

[120] Wooten, M. W, et al. Essential role of sequestosome 1/in regulating accumulation of Lys63-ubiquitinated proteins. The Journal of biological chemistry, (2008). p. 6783-9., 62.

[121] Ikeda, F, & Dikic, I. Atypical ubiquitin chains: new molecular signals.'Protein Modifications: Beyond the Usual Suspects' review series. EMBO reports, (2008). , 536-542.

[122] Johansen, T, & Lamark, T. Selective autophagy mediated by autophagic adapter proteins. Autophagy, (2011). , 279-296.

[123] Nezis, I. P. Selective autophagy in Drosophila. International journal of cell biology, (2012). , 146767.

[124] Nezis, I. P, et al. Ref(2)P, the Drosophila melanogaster homologue of mammalian is required for the formation of protein aggregates in adult brain. The Journal of cell biology, (2008). p. 1065-71., 62.

[125] Filimonenko, M, et al. The selective macroautophagic degradation of aggregated proteins requires the PI3P-binding protein Alfy. Molecular cell, (2010). , 265-279.

[126] Simonsen, A, et al. Alfy, a novel FYVE-domain-containing protein associated with protein granules and autophagic membranes. Journal of cell science, (2004). Pt 18): , 4239-4251.

[127] Clausen, T. H, et al. SQSTM1 and ALFY interact to facilitate the formation of p62 bodies/ALIS and their degradation by autophagy. Autophagy, (2010). p. 330-44., 62.

[128] Fujita, N, et al. The Atg16L complex specifies the site of LC3 lipidation for membrane biogenesis in autophagy. Molecular biology of the cell, (2008). , 2092-2100.

[129] Finley, K. D, et al. blue cheese mutations define a novel, conserved gene involved in progressive neural degeneration. The Journal of neuroscience : the official journal of the Society for Neuroscience, (2003). , 1254-1264.

Permissions

The contributors of this book come from diverse backgrounds, making this book a truly international effort. This book will bring forth new frontiers with its revolutionizing research information and detailed analysis of the nascent developments around the world.

We would like to thank Yannick Bailly, for lending his expertise to make the book truly unique. He has played a crucial role in the development of this book. Without his invaluable contribution this book wouldn't have been possible. He has made vital efforts to compile up to date information on the varied aspects of this subject to make this book a valuable addition to the collection of many professionals and students.

This book was conceptualized with the vision of imparting up-to-date information and advanced data in this field. To ensure the same, a matchless editorial board was set up. Every individual on the board went through rigorous rounds of assessment to prove their worth. After which they invested a large part of their time researching and compiling the most relevant data for our readers. Conferences and sessions were held from time to time between the editorial board and the contributing authors to present the data in the most comprehensible form. The editorial team has worked tirelessly to provide valuable and valid information to help people across the globe.

Every chapter published in this book has been scrutinized by our experts. Their significance has been extensively debated. The topics covered herein carry significant findings which will fuel the growth of the discipline. They may even be implemented as practical applications or may be referred to as a beginning point for another development. Chapters in this book were first published by InTech; hereby published with permission under the Creative Commons Attribution License or equivalent.

The editorial board has been involved in producing this book since its inception. They have spent rigorous hours researching and exploring the diverse topics which have resulted in the successful publishing of this book. They have passed on their knowledge of decades through this book. To expedite this challenging task, the publisher supported the team at every step. A small team of assistant editors was also appointed to further simplify the editing procedure and attain best results for the readers.

Our editorial team has been hand-picked from every corner of the world. Their multi-ethnicity adds dynamic inputs to the discussions which result in innovative outcomes. These outcomes are then further discussed with the researchers and contributors who give their valuable feedback and opinion regarding the same. The feedback is then collaborated with the researches and they are edited in a comprehensive manner to aid the understanding of the subject.

Apart from the editorial board, the designing team has also invested a significant amount of their time in understanding the subject and creating the most relevant covers. They scrutinized every image to scout for the most suitable representation of the subject and create an appropriate cover for the book.

The publishing team has been involved in this book since its early stages. They were actively engaged in every process, be it collecting the data, connecting with the contributors or procuring relevant information. The team has been an ardent support to the editorial, designing and production team. Their endless efforts to recruit the best for this project, has resulted in the accomplishment of this book. They are a veteran in the field of academics and their pool of knowledge is as vast as their experience in printing. Their expertise and guidance has proved useful at every step. Their uncompromising quality standards have made this book an exceptional effort. Their encouragement from time to time has been an inspiration for everyone.

The publisher and the editorial board hope that this book will prove to be a valuable piece of knowledge for researchers, students, practitioners and scholars across the globe.

List of Contributors

Thabata Lopes Alberto Duque
Laboratory of Cell Biology, Oswaldo Cruz Institute, Oswaldo Cruz Foundation, Rio de Janeiro, RJ, Brazil Laboratory of Cell Biology, Department of Biology, Federal University of Juiz de Fora, MG, Brazil

Xênia Macedo Souto
Laboratory of Cell Biology, Oswaldo Cruz Institute, Oswaldo Cruz Foundation, Rio de Janeiro, RJ, Brazil Laboratory of Structural Biology, Oswaldo Cruz Institute, Oswaldo Cruz Foundation, Rio de Janeiro, RJ, Brazil

Valter Viana de Andrade-Neto
Laboratory of Biochemistry of Trypanosomatids, Oswaldo Cruz Institute, Oswaldo Cruz Foundation, Rio de Janeiro, RJ, Brazil

Vítor Ennes-Vidal
Laboratory of Molecular Biology and Endemic Diseases, Oswaldo Cruz Institute, Oswaldo Cruz Foundation, Rio de Janeiro, RJ, Brazil

Rubem Figueiredo Sadok Menna-Barreto
Laboratory of Cell Biology, Oswaldo Cruz Institute, Oswaldo Cruz Foundation, Rio de Janeiro, RJ, Brazil

Aiguo Wu
Department of Molecular Microbiology, George Mason University, Manassas VA, USA

Yian Kim Tan
DSO National Laboratories, Singapore, Singapore

Hao A. Vu
Biosecurity Research Institute, Kansas State University, Manhattan, USA

Patricia Silvia Romano
Laboratory of Trypanosoma cruzi and the Host Cell, Institute of Histology and Embryology, National Council of Scientific and Technical Research (IHEM-CONICET), School of Medicine-National University of Cuyo, Mendoza, Argentina

Nikolai V. Gorbunov, Pei-Jyun Liao and Min Zhai
The Henry M. Jackson Foundation, USA

Thomas B. Elliott
Armed Forces Radiobiology Research Institute, USA

Dennis P. McDaniel and K. Lund
School of Medicine, Uniformed Services University of the Health Sciences, Bethesda, Maryland, USA

Juliann G. Kiang
Armed Forces Radiobiology Research Institute, USA

School of Medicine, Uniformed Services University of the Health Sciences, Bethesda, Maryland, USA

Audrey Ragagnin, Aurélie Guillemain and Yannick J. R. Bailly
Cytologie & Cytopathologie Neuronales, INCI CNRS UPR3212, Université de Strasbourg, Strasbourg, France

Nancy J. Grant
Trafic Membranaire Dans les Cellules Neurosécrétrices et Neuroimmunitaires, INCI CNRS UPR3212, Université de Strasbourg, Strasbourg, France

Kah-Leong Lim
Department of Physiology, National University of Singapore, Singapore National Neuroscience Institute, Singapore Duke-NUS Graduate Medical School, Singapore

Grace G.Y. Lim
Department of Physiology, National University of Singapore, Singapore

Chengwu Zhang
National Neuroscience Institute, Singapore

Malgorzata Gajewska, Katarzyna Zielniok and Tomasz Motyl
Department of Physiological Sciences, Faculty of Veterinary Medicine, Warsaw University of Life Sciences, Warsaw, Poland

M.L. Escobar, O.M. Echeverría and G.H. Vázquez-Nin
Departamento de Biología Celular, Facultad de Ciencias. Universidad Nacional Autónoma de México, Mexico

Cindy K. Miranti
Laboratory of Integrin Signaling and Tumorigensis, Van Andel Research Institute, Grand Rapids, Michigan, USA

Eric A. Nollet
Laboratory of Integrin Signaling and Tumorigensis, Van Andel Research Institute, Grand Rapids, Michigan, USA Van Andel Institute Graduate School, Grand Rapids, Michigan, USA

Sebastian Wolfgang Schultz and Andreas Brech
Department of Biochemistry, Institute for Cancer Research, Oslo University Hospital, The Norwegian Radium Hospital, Oslo, Norway Centre for Cancer Biomedicine, Faculty of Medicine, University of Oslo, Oslo, Norway

Ioannis P. Nezis
School of Life Sciences, University of Warwick, Coventry, United Kingdom